AF598321

The Kelvin Problem

Lord Kelvin, as portrayed by Vanity Fair.

The Kelvin Problem

Foam Structures of Minimal Surface Area

EDITED BY

DENIS WEAIRE
TRINITY COLLEGE, DUBLIN, IRELAND

UK Taylor & Francis Ltd, 1 Gunpowder Square, London EC4A 3DE
USA Taylor & Francis Inc., 1900 Frost Road, Suite 101, Bristol, PA 19007

British Library Cataloguing in Publication Data

A catalogue record for this book is available from the British Library.
ISBN 0 7484 06328 (cased)

Library of Congress Cataloguing in Publication data are available

Cover design by Youngs Design in Production.

Typeset by The Studio Design & Typesetting, Exeter, Devon.

Printed in Great Britain by T.J. International, Padstow, Cornwall.

Contents

Preface

The present search for surface-minimising structures has given new significance to the Frank-Kasper structures. These have turned out to be the best candidates, when converted by the dual transformation *point* (*atom*) → *cell.* I have therefore been asked how these structures first arose in my own work.

Together with John Kasper, who was well acquainted with some complex alloy structures from their experimental determination, I set out to put some system into their classification. We had both visited Toledo earlier in the year that we met, and both had been struck by the same idea, that Moorish tiling patterns played on the same theme, attempting the impossible, the tiling of the plane with regular pentagons. Some walls in Toledo seemed to accomplish this, up to a boundary, beyond which you could not continue; and a dome in the synagogue succeeds, but on a spherically curved surface. Elsewhere, the tiling with equilateral pentagons is found. A floor in the Glasgow Physics Lab. also shows this pattern, or used to.

Then, between us, we perceived that some alloy structures can be seen as alternate stackings of the equilateral pentagon outline and its dual. Adding in some elementary topology from Euler (inspired, so far as we were concerned, by Cyril Smith), we had a publication. Then I invited John Kasper to come to Bristol for a year, and we wrote another.

In comparing this with recent developments, two thoughts emerge: that in thinking about structures, inspiration can be drawn from surprising sources, and that Cyril Stanley Smith's sowed many of the seeds of such eclectic thought, over more than one generation.

Charles FRANK

Acknowledgements

This work was first published as a special issue of *Forma* and is published in its present form, with minor corrections and alterations, by agreement with KTC Scientific Publishers (Tokyo). The excerpts from Lord Kelvin's notebooks are from the Kelvin Collection, University of Cambridge, and are reproduced by kind permission of the University Library. The photograph of Kelvin's wire model is reproduced courtesy of the University of Glasgow. "Close-packing and Froth" is reprinted by kind permission of the Illinois Journal of Mathematics. The portraits of Kelvin, Plateau and Smith are by Hugh Masterson.

We are especially grateful to Sir Charles Frank for agreeing to contribute a Preface. The special issue was compiled under the general editorship of R. Takaki.

Note

We are sad to report the untimely death of Fred Almgren on February 5, 1997. As one of the pioneers of geometric measure theory, Almgren was responsible for the basic definitions and existence results for mathematical clusters of soap bubbles, as reported in his contribution to this volume. As one of the founders of the Geometry Center in Minneapolis, Almgren led early efforts to study geometry with computers, rekindling interest in Kelvin's problem among mathematicians, and directly inspiring the development of the software later used by Weaire and Phelan to discover their new example.

Introduction

D. WEAIRE

Department of Pure and Applied Physics, Trinity College, Dublin 2, Ireland

The Kelvin Problem

The Kelvin Problem is that of the partitioning of three-dimensional space into cells of equal volume and minimum surface area. It is one of those well-defined, rather abstract questions that rise out of the complexity and confusion of scientific research as a focus of attention. Like many others, it is easily posed but not easily solved. Even if we beat a strategic retreat into two dimensions we find, amazingly, that no proof exists to support the general conviction that the bee's honeycomb minimises the line length of equal cells. In three dimensions we cannot even be sure what the solution should be, let alone confer upon it the full mathematical respectability of a rigorous proof.

For a hundred years Kelvin's proposed solution resisted challenges, in the form of various rival structures. It also failed to appear in experiments. In 1994 Robert Phelan, in his first term as a research student, and myself produced the first successful challenge. The recipe was one third intuition, one third analogy, and one third serendipidity. Since then there have been attempts to follow that lead and discover related structures which might fill space yet more efficiently. So far they have not done so.

In this volume we have gathered together some papers representative of the history of the problem and the current contributions to it, mostly by mathematicians and physicists. We include only a little of the work of Plateau, which guided Kelvin. That work is particularly impressive when it is realised that Plateau became blind in 1843 as the result of an optical experiment. Despite his disability he found and enunciated the rules of equilibrium of soap films. Kelvin was to use these since, as he says in opening his paper "the problem is solved in foam". This is simply because the films which constitute the foam have energy proportional to surface area. The rules narrowed the range of possibilities which he had to consider, once he had chosen to restrict attention to identical cells with identical orientation. Note that the general problem makes no such restriction at the outset: hence the occasional confusion as to what the problem is supposed to be, and the validity of Kelvin's solution.

No-one can tell for how much longer this will remain an open problem. Computational proofs present a new style of mathematics which may eventually dispose of it, but there is little sign of progress. In the meantime we have powerful simulation techniques to test educated guesses. The fact that so little is cut-and-dried is part of the fascination of the field.

I would like to mention the influence of Cyril Stanley Smith, who died in 1992. Although he mainly rejoiced in structures which had some interplay of order and disorder, he was not averse to the contemplation of ordered beauty as well. The eloquent expression of his tantalising and tentative visions of the principles of form, for example in the compilation *A Search for Structure*, inspired many to become what he called *philomorphs*.

Kelvin on Foam

The seminal contribution of Lord Kelvin (then Sir William Thomson) represented only a brief episode in his career, and is hardly mentioned in the many comprehensive biographies of the great man. He published one important paper directly addressed to what we have called the Kelvin Problem. It was little more than a temporary diversion from the main theme of his ideas about ether, which centred on his celebrated vortex model. In addition to his suggestion of a solution to the problem of dividing three dimensional space most economically, in terms of area, he made various cryptic remarks about foam elasticity in subsequent papers. This published work leaves much unsaid, regarding his reasoning. Recourse may be made to his notes, to better understand all of this, with limited success.

Kelvin kept a remarkable series of notebooks, in which he jotted down his thoughts at all times of the day and in whatever location he found himself. Untidy, rambling and fragmentary, they present a splendid record of the thought processes of a theoretical scientist, by turns confused, frustrated or elated.

The idea that foam might be relevant in the context of mechanical models of the ether seems to have occurred to him in late September. His first notes on it are dated Sept. 20, 1887. He lay in bed in the early morning, thinking about the rigidity of foam. By the following day he had posed for himself the mathematical problem which arises when one considers the ideal structure of foam made up of equal bubbles. It was a peculiar sort of foam that he envisaged—one with no gas enclosed, and hence requiring some sort of retaining boundary.

It is clear that he had at his side Plateau's classic work on soap films, since he noted page numbers at key points. He set out to find a structure which obeyed Plateau's rules. By November 4, he had found it, after toying with some other possibilities. From the outset, he called the cell a *tetrakaidecahedron*, a name that some have found unpleasant: its antecedents are unclear. Surprisingly perhaps, he seems to have come upon this form of cell by a process of construction of the individual polyhedron, rather than by first constructing the Voronoi (or Wigner-Seitz) cell of the body-centred cubic lattice. Kelvin was expert in crystallography (to such an extent that Heaviside said his brain had crystallised), so it might have been thought that he took the latter route.

In the published paper he pursues quite a tortuous path to the proposed structure. First he notes Plateau's demonstration of the instability of the multiple vertex formed by soap films at the centre of a cubical wire frame. He observes that the unstable configuration may be viewed as part of the structure obtained by packing rhombic dodecahedra. If the same rearrangement is imposed on all such vertices, the structure reverts to one of tetrakaidecahedra, but requires to be strained to make it perfectly symmetric. All of this is much clearer from the modern standpoint: the first structure has the face-centred cubic lattice. With the rearrangements it becomes face-centred tetragonal. This lattice and the body-centred tetragonal lattice are equivalent. Kelvin's subsequent adjustment simply restores cubic symmetry in the body-centred cubic lattice.

He proceeded to rapidly calculate the subtle distortions needed to bring this cell into complete conformity with Plateau's rules, that is, to give it the correct angles of intersection while maintaining zero mean surface curvature. By November 4 he was already noting a possible title of his proposed paper: "The Division of Space into Equal Volumes of Least Possible Area". In due course it was to become "On the Division of Space with Minimum Partitional Area".

The paper was published in *Philosophical Magazine* in the following month. This speed of publication is fairly typical of the time and calls into question any notion of progress in the publishing industry. It may have helped that he was the Editor. Although well into his sixties, he was still close to the peak of his productivity: of his 660 papers, more than twenty appeared in that year.

We have found no mention in the notebooks of any attempt by Kelvin to look for the ideal structure himself in the extended structure of a real foam. This might be expected, since he was a practically minded scientist. Indeed he had a wire model made of his structure, subsequently nicknamed "Kelvin's Bedspring". It was possible to introduce soap films into this, but this begs most of the important experimental questions.

It must be remembered that his original interest was in the nature of light and the "all-pervading ether" of space which fascinated the natural philosophers of that century. This is by no means the only example of a contribution by Kelvin to materials science motivated in this way: his demonstration of the properties of a highly viscous liquid, in the form of pitch, is well known.

His choice of a foam model for the ether was motivated by its elastic properties, in relation to wave propagation. He convinced himself that a foam without gas, retained by some sort of boundary to prevent collapse, would have a zero elastic modulus corresponding to stretching or compression in one direction only. This neatly eliminates the longitudinal elastic waves which were the undesirable feature of the modelling of the ether as an elastic medium: only transverse light waves are observed. It is in itself a remarkable insight. Only now, as large computations are used to analyse foam mechanics, can we clearly test its validity. The modulus is indeed *approximately* zero. In some places Kelvin covered himself by saying it was only an approximation. In others he claimed it to be an exact result, without ever clearly explaining how he arrived at it.

It turns out that the statement is *exactly* true for the much simpler equivalent problem in two dimensions. In that case we deal with the hexagonal honeycomb as the ideal structure. Its sides can be adjusted, keeping its angles fixed, in such a way that it can be stretched in one direction with an exactly linear increase in total line length (i.e. energy). Hence there is a zero modulus (second derivative). Intriguingly, Kelvin's notebooks have many diagrams of the honeycomb. The 1887 paper uses it to explain vertex instability, but in general he probably drew much wider inspiration from it than is evident from the paper.

It seems that he developed his ideas on elasticity in two dimensions and relied on an imprecise generalisation to three dimensions.

To those who have in recent years founded their approach to three-dimensional foams on the study of their two-dimensional counterparts, such an approach is familiar. What is astonishing is the soundness of his intuition and the rapidity with which he developed it into a full theory, with minimal inputs or validation from others.

At the outset he dismissed the notion that a material with a negative bulk modulus would be unstable. Provided the boundary was fixed, he claimed to be able to prove stability. However, this argument is certainly wrong. The honeycomb structure is metastable and the Kelvin structure unstable with respect to exchange of area/volume between cells, keeping the total constant. In a real foam such a mechanism could only be exercised by the slow diffusion of gas through the soap films. This rather contradicts a remark of C. S. Smith, to the effect that even if surpassed in terms of area minimisation, Kelvin's structure would remain the only one stable against diffusion. Not so. The point is that the cell pressures must be equal, since all cells are equivalent, but this only implies that the the perfect structure suffers no diffusion: it is unstable with respect to infinitesimal fluctuations of cell volume. Remember that Kelvin's foam model had no gas, so that it must collapse instantaneously on account of such an instability.

There thus emerged the idea of the ether as a foam. Stoney, in his preface to Fournier d'Albe's *The Electron Theory*, described this as an ancient idea but I am aware of no earlier reference to it. It seems to have generated no great immediate interest, in spite of Kelvin's standing. Probably increasing doubts about stability caused Kelvin's retreat from the model. He remained interested in material ether models until his death. His continued insistence on them became increasingly outdated by the more abstract electromagnetic theory of Maxwell, promulgated by Fitzgerald and others. Thus his brilliant ideas on foam structure and mechanics lay disregarded, until later generations of scientists from many fields found cause to return to them.

It was D'Arcy Wentworth Thompson who gave Kelvin's proposition more prominence by discussing it extensively in his classic work "On Growth and Form". This was influential among biologists. His propaganda on behalf of the tetrakaidecahedron was the subject of some irony in the comments of Bonner, as editor of a later edition. He based his skepticism on the work of Matzke, who searched unsuccessfully for Kelvin's tetrakaidecahedra in a real foam. Little or no trace of his structure could be found in foams or analogous structures in nature. And the search for a proof of his conjecture also proved fruitless.

Portraits of Plateau and Smith

Joseph Antoine Ferdinand Plateau.

Cyril Stanley Smith.

Plateau's Wire Frame Experiments

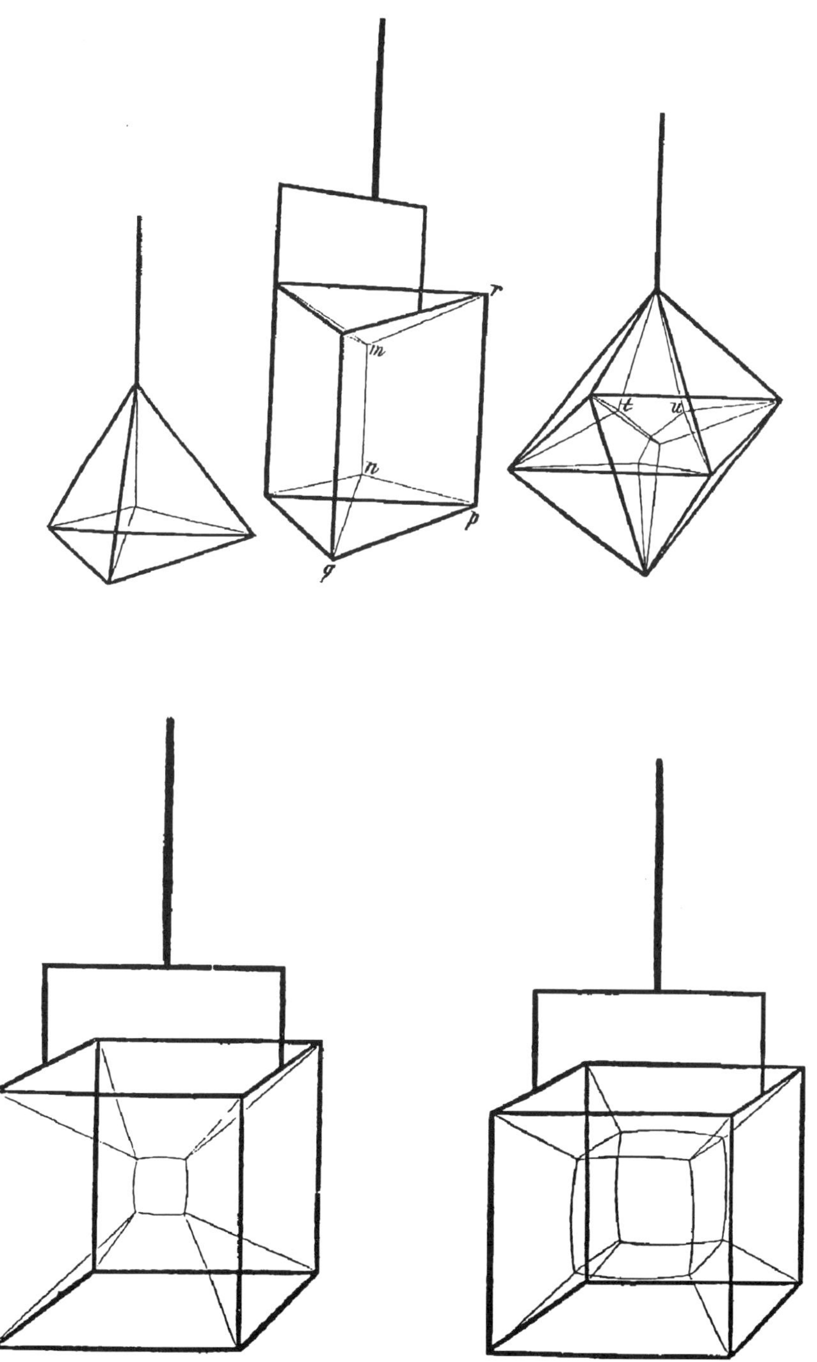

Some Quotations and Excerpts

selected by D. WEAIRE

Joseph Antoine Ferdinand Plateau[1]

1873, translated from the French

188. The two laws which I have just discussed must now, I think, be regarded as well established for all laminar assemblages. Now, these laws lead us to a very remarkable consequence: the froth which forms on certain liquids, for example on champagne, beer or agitated soap water, is evidently an assemblage of liquid films, composed of many small films or partitions intersecting each other and enclosing small amounts of gas. Consequently, though everything in the froth seems ruled by chance, it must be subject to those same laws; thus its innumerable partitions necessarily join each other three-by-three at equal angles, and all its edges are distributed in such a way that there are always four of them meeting at the same point, making equal angles between them.

James Clerk Maxwell[2]

1874

Here, for instance, we have a book, in two volumes, octavo, written by a distinguished man of science, and occupied for the most part with the theory and practice of bubble-blowing. Can the poetry of bubbles survive this?

Which, now, is the more poetical idea—the Etruscan boy blowing bubbles for himself, or the blind man of science teaching his friends to blow them, and making out by a tedious process of question and answer the condition of the forms and tints which he can never see?

Kelvin (Notebooks)[3]

Sept. 29, 1887, in bed.

Rigidity of foam (accompanied by drawings of honeycomb).

Sept. 30, 1887

Problem Divide Space into equal volumes, by minimum area.

Nov. 4, 1887, 6.45 pm.

Truncate the corners of a regular octahedron to such extent that the side of each square is equal to the residue of each side. This makes a tetrakaidecahedron. [...] Space can be filled by equal and similar figures placed together. Now curve each edge; those of squares convexly in their own planes, [...]. The Division of space into equal volumes of least possible partitional area.

Agnes G. King (niece of Lord Kelvin)[4]

Nov. 5, 1887

When I arrived here yesterday Uncle William and Aunt Fanny met me at the door, Uncle William armed with a vessel of soap and glycerine prepared for blowing soap bubbles, and a tray with a number of mathematical figures made of wire. These he dips into the soap mixture and a film forms or adheres to the wires very beautifully and perfectly regularly. With some scientific end in view he is studying these films.

Kelvin to Rayleigh[3]

Nov. 20, 1887

I have become involved in another affair [...] which George Darwin characterises as utterly frothy.

George Francis Fitzgerald to Kelvin[3]

1896

··· a pure waste of time.

D'Arcy Wentworth Thompson[5]

1917

Moreover, the perfection of mathematical beauty is such (as Colin Maclaurin learned of the bee), that whatsoever is most beautiful and regular is also found to be most useful and excellent.

Hermann Weyl[6]

1952

I am inclined to believe that Lord Kelvin's configuration gives the absolute minimum; but so far as I know, this has never been proved.

Fred J. Almgren[7]

1982

Despite the claims of various authors to the contrary it seems an open question whether or not Lord Kelvin's candidate is optimal or even whether any optimal partitioning could be periodic.

REFERENCES

1. J. A. F. Plateau, Statique expérimentale et théorique des liquides soumis aux seules forces moléculaires (Gauthier-Villars, Paris) (1873).
2. J. C. Maxwell, Nature, **10**, 119 (1874).
3. Kelvin Correspondence, Cambridge University Library.
4. A. G. King, Kelvin the Man (Hodder and Stoughton, London) 192 (1925).
5. D'A. W. Thompson, On Growth and Form, abridged edition (ed. J. T. Bonner) (Cambridge University Press, Cambridge) (1961) (First edition 1917).
6. H. Weyl, Symmetry (Princeton University Press, Princeton) (1952).
7. F. J. Almgren, Jr, The Mathematical Intelligencer, **4**, 164 (1982).

Portrait of Kelvin

Sir William Thomson, Lord Kelvin.

Kelvin's Wire Model

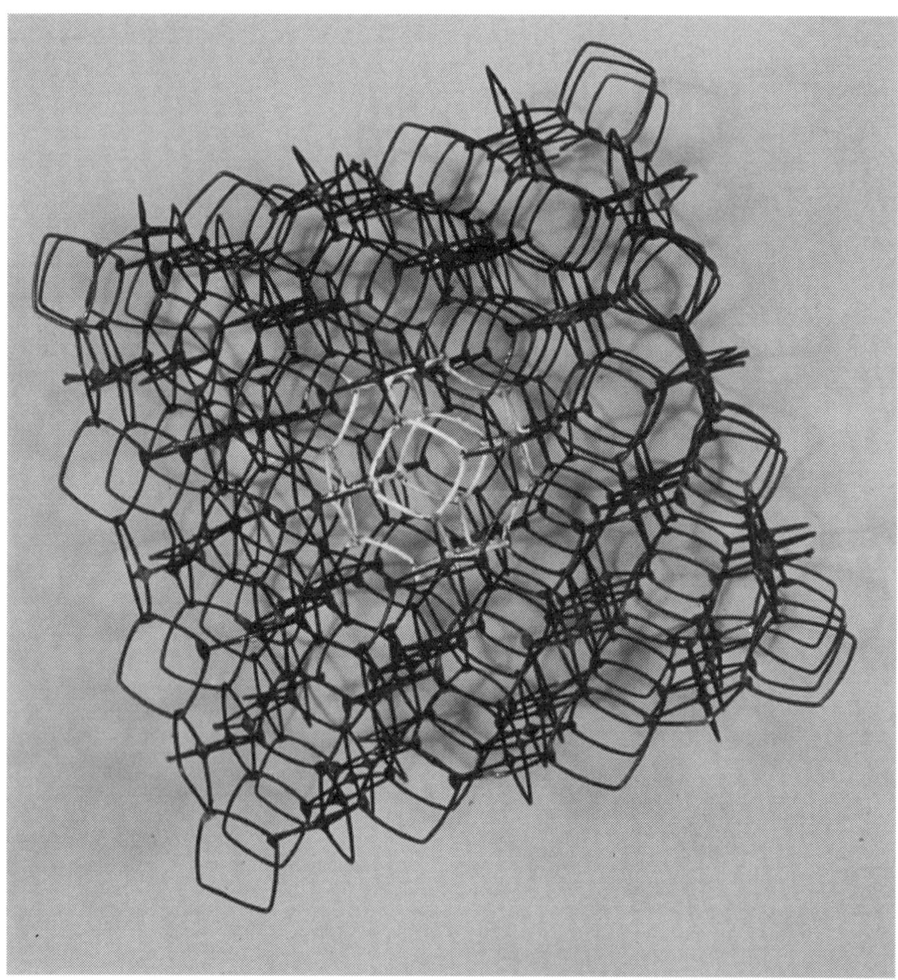

Kelvin's wire model (courtesy of the University of Glasgow).

Some Pages from the Notebooks of Kelvin*

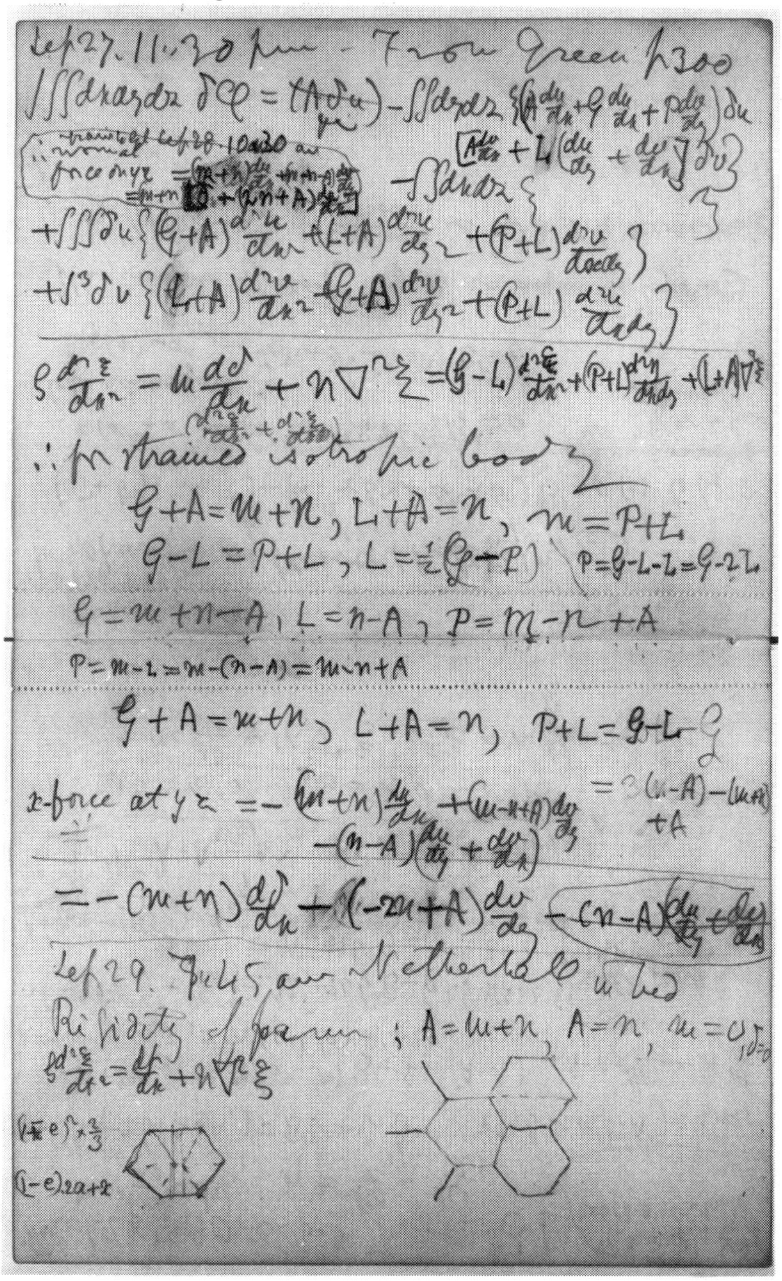
Sep 27. 11.30 pm – From Green p 300

$\mathfrak{g} + A = m + n,\ L + A = n,\ m = P + L$

$\mathfrak{g} - L = P + L,\ L = \frac{1}{2}(\mathfrak{g} - P)\qquad P = \mathfrak{g} - L - L = \mathfrak{g} - 2L$

$\mathfrak{g} = m + n - A,\ L = n - A,\ P = m - n + A$

$P = m - L = m - (n - A) = m - n + A$

$\mathfrak{g} + A = m + n,\ L + A = n,\ P + L = \mathfrak{g} - L$

Sep 29. 9.45 am Netherhall in bed

1

*Courtesy of Cambridge University Library.

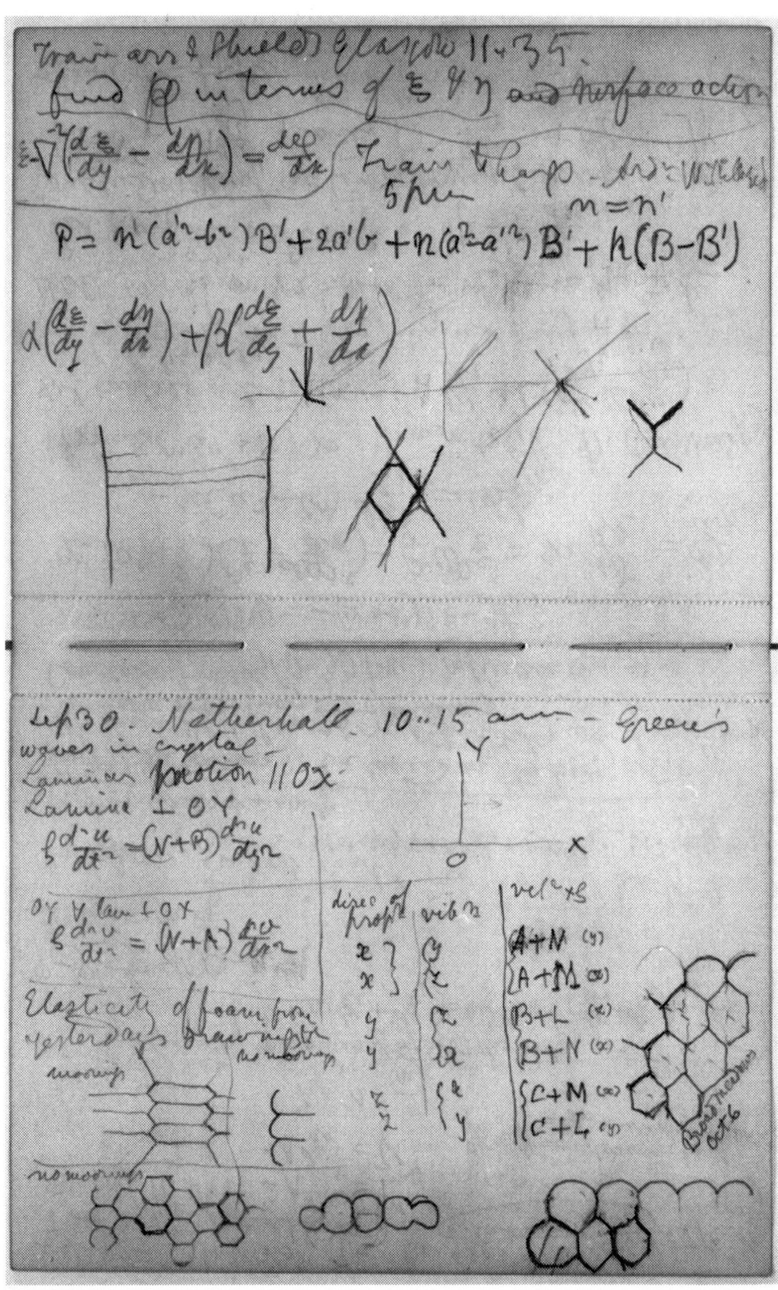

2

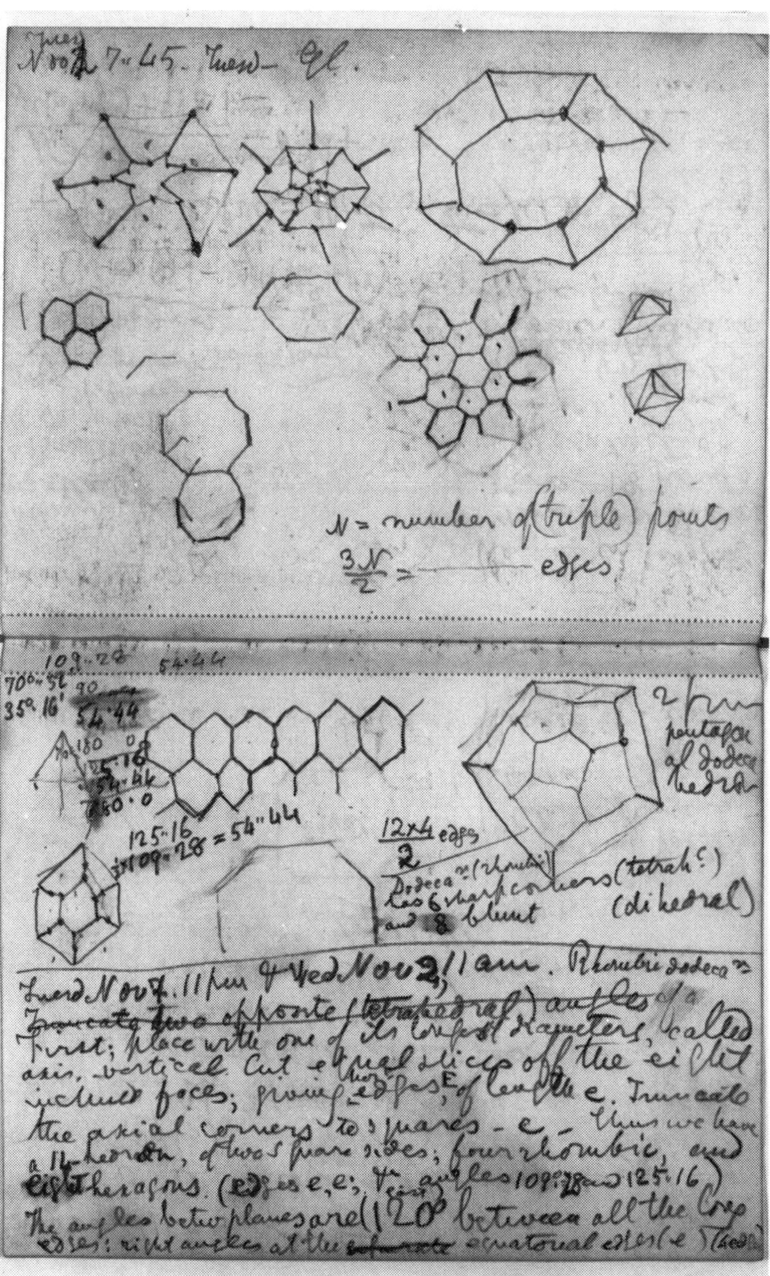

3

Truncate the corners of a regular octahedron to such extent that the side of each square is equal to the residue of each side. This makes a tetrakaidecahedron six square sides, and 8 regular hexagons; all its 36 edges are equal, and space can be filled by equal and similar figures placed [illegible]

Now bend all the edges [illegible] curve each edge; those of squares convexly in their own planes, of the others concavely in the planes bisecting the planes of the hexagons.

The division of space into equal volumes of least possible partitional area is by t[illegible]ns, with curved edges, and minimal surfaces

11h30 pm, for tetrakai $= \frac{\frac{\sqrt{2}}{3}\left(1-\frac{4}{27}\right)}{2\sqrt{3}\left(1-\frac{4}{9}\right)+8.\frac{1}{9}}$

$= \frac{1}{3\sqrt{6}} \cdot \frac{1-4/27}{1-4/9.\left(1-\frac{1}{\sqrt{3}}\right)}$

Proper discriminant is this line divided by $\sqrt[3]{vol}$

Sat Nov 5. Gl. 2h30 pm — 7: $(x^3+3xy^2)h$ carriage to Garscube

[illegible]

4

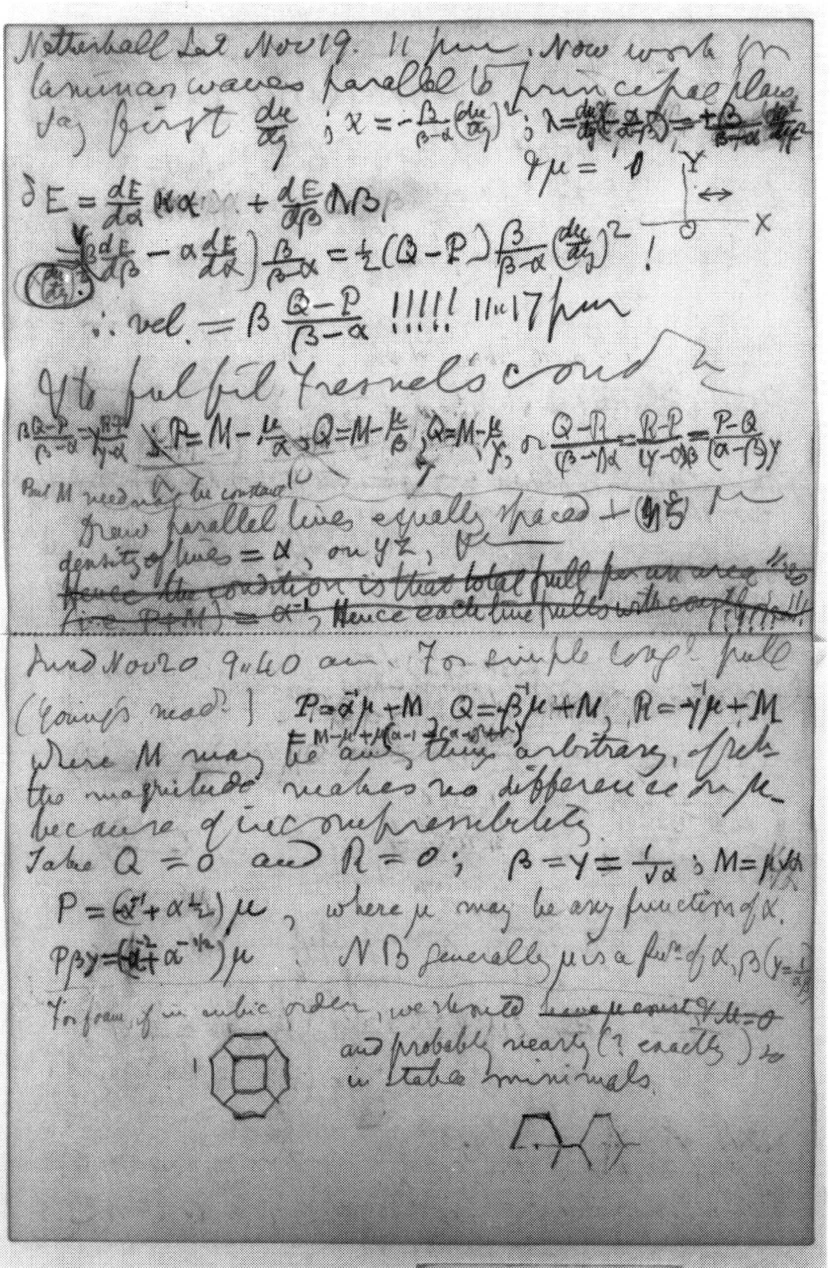

5

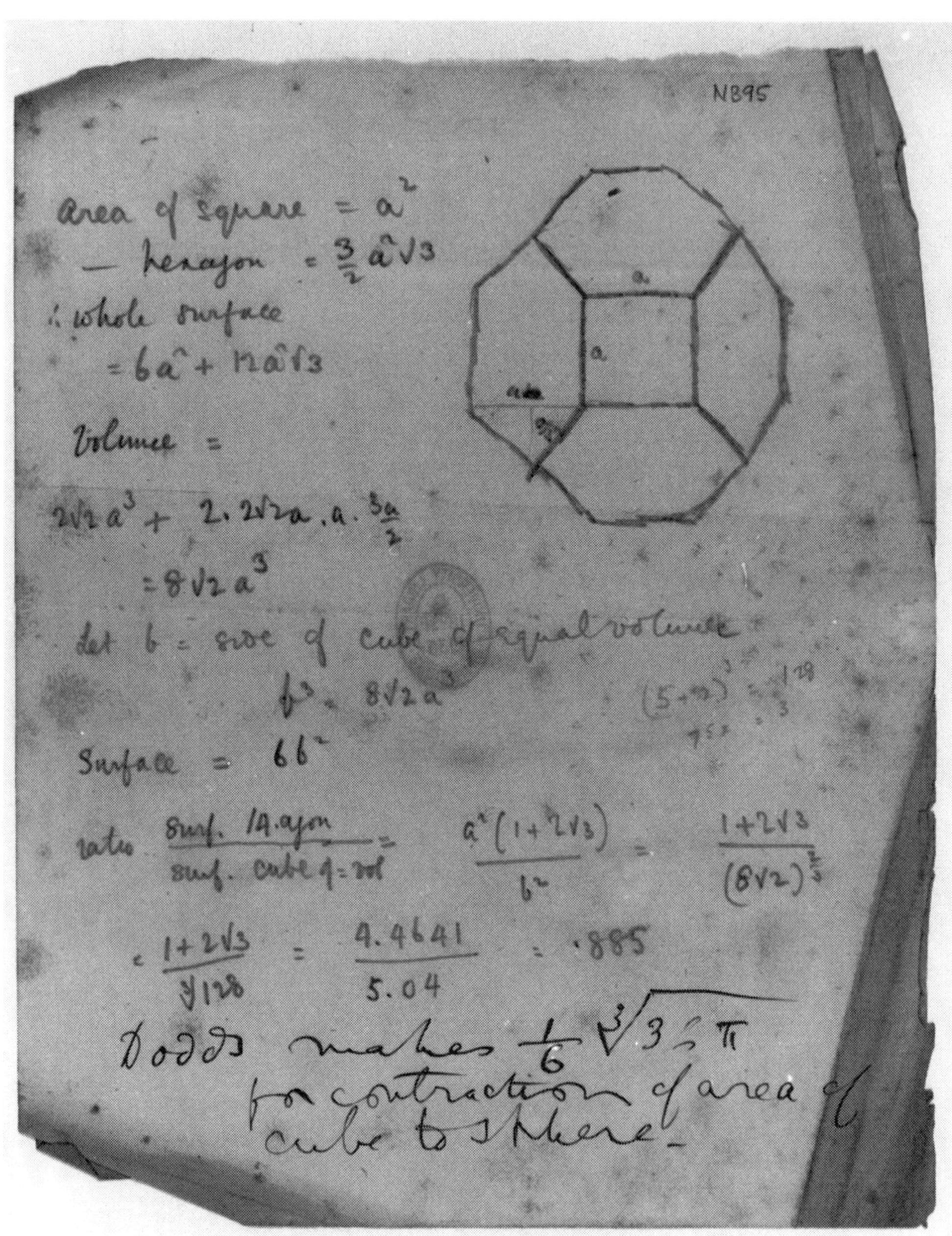

N895

area of square $= a^2$

— hexagon $= \frac{3}{2}a^2\sqrt{3}$

$\therefore$ whole surface

$= 6a^2 + 12a^2\sqrt{3}$

Volume =

$2\sqrt{2}\,a^3 + 2 \cdot 2\sqrt{2}a \cdot a \cdot \frac{3a}{2}$

$= 8\sqrt{2}\,a^3$

Let b = side of cube of equal volume

$b^3 = 8\sqrt{2}\,a^3$

Surface $= 6b^2$

ratio $\frac{\text{surf. 14-agon}}{\text{surf. cube of = vol}} = \frac{a^2(1+2\sqrt{3})}{b^2} = \frac{1+2\sqrt{3}}{(8\sqrt{2})^{\frac{2}{3}}}$

$= \frac{1+2\sqrt{3}}{\sqrt[3]{128}} = \frac{4.4641}{5.04} = \cdot 885$

Dodds makes $\frac{1}{6}\sqrt[3]{36\pi}$ for contraction of area of cube to sphere.

6

On the Division of Space with Minimum Partitional Area*

Sir W. THOMSON

1. This problem is solved in foam, and the solution is interestingly seen in the multitude of film-enclosed cells obtained by blowing air through a tube into the middle of a soap-solution in a large open vessel. I have been led to it by endeavours to understand, and to illustrate, Green's theory of "extraneous pressure" which gives, for light traversing a crystal, Fresnel's wave-surface, with Fresnel's supposition (strongly supported as it is by Stokes and Rayleigh) of velocity of propagation dependent, not on the distortion-normal, but on the line of vibration. It has been admirably illustrated, and some elements towards its solution beautifully realized in a manner convenient for study and instruction, by Plateau, in the first volume of his *Statique des Liquides soumis aux seules Forces Moléculaires*.

2. The general mathematical solution, as is well known, is that every interface between cells must have constant curvature** throughout, and that where three or more interfaces meet in a curve or straight line their tangent-planes through any point of the line of meeting intersect at angles such that equal forces in these planes, perpendicular to their line of intersection, balance. The *minimax* problem would allow any number of interfaces to meet in a line; but for a pure minimum it is obvious that not more than three can meet in a line, and that therefore, in the realization by the soap-film, the equilibrium is necessarily unstable if four or more surfaces meet in a line. This theoretical conclusion is amply confirmed by observation, as we see at every intersection of films, whether interfacial in the interior of groups of soap-bubbles, large or small, or at the outer bounding-surface of a group, never more than three films, but, wherever there is intersection, always *just three films*, meeting in a line. The theoretical conclusion as to the angles for stable equilibrium (or pure minimum solution of the mathematical problem) therefore becomes, simply, that every angle of meeting of film-surfaces is exactly 120°.

3. The rhombic dodecahedron is a polyhedron of plane sides between which every angle of meeting is 120°; and space can be filled with (or divided into) equal and similar rhombic dodecahedrons. Hence it might seem that the rhombic dodecahedron is the

---

*Reproduced from *Phil. Mag.* (1887) Vol. **24**, No. 151, p. 503.

Keywords: Foam, Structure, Minimum Partition Area, Tetrakaidecahedron, Space Division (given by the editors of FORMA).

**By "curvature" of a surface I mean sum of curvatures in mutually perpendicular normal sections at any point; not Gauss' "curvatura integra", which is the product of the curvature in the two "principal normal sections", or sections of greatest and least curvature. (See Thomson and Tait's "Natural Philosophy", part i, §130 and §136.)

solution of our problem for the case of all the cells equal in volume, and every part of the boundary of the group either infinitely distant from the place considered, or so adjusted as not to interfere with the homogeneousness of the interior distribution of cells. Certainly the rhombic dodecahedron *is a solution of the minimax, or equilibrium-problem*; and certain it is that no other plane-sided polyhedron can be a solution.

4. But is has seemed to me, on purely theoretical consideration, that the tetrahedral angles of the rhombic dodecahedron*, giving, when space is divided into such figures, twelve plane films meeting in a point (as twelve planes from the twelve edges of a cube meeting in the centre of the cube) are essentially unstable. That it is so is proved experimentally by Plateau (vol. i, §182, Fig. 71) in his well-known beautiful experiment with his cubic skeleton frame dipped in soap-solution and taken out. His Fig. 71 is reproduced here in Fig. 1. Instead of twelve *plane* films stretched inwards from the twelve edges and meeting in the centre of the cube, it shows twelve films, of which eight are slightly curved and four are plane**, stretched from the twelve edges to a small central plane quadrilateral film with equal curved edges and four angles each of 109°28′. Each of the plane films is an isosceles triangle with two equal curved sides meeting at a corner of the central curvilinear square in a plane perpendicular to its plane. It is in the plane through an edge and the centre of the cube. The angles of this plane curvilinear triangle are respectively 109°28′, at the point of meeting of the two curvilinear sides: and each of the two others half of this, or 54°44′.

5. I find that by blowing gently upon the Plateau cube into any one of the square apertures through which the little central quadrilateral film is seen as a line, this film is caused to contract. If I stop blowing before it contracts to a point, it springs back to its primitive size and shape. If I blow still very gently but for a little more time, the quadrilateral contracts to a point, and the twelve films meeting in it immediately draw out a fresh little quadrilateral film similar to the former, but in a plane perpendicular to its plane and to the direction of the blast. Thus, again and again, may the films be transformed so as to render the little central curvilinear square parallel to one or other of the three pairs of square apertures of the cubic frame. Thus we see that the twelve plane films meeting in the centre of the cube is a configuration of unstable equilibrium which may be fallen from in three different ways.

---

*The rhombic dodecahedron has six tetrahedral angles and eight trihedral angles. At each tetrahedral angle the plane faces cut one another successively at 120°, while each is perpendicular to the one remote from it; and the angle between successive edges is $\cos^{-1}(1/3)$, or 70°32′. The obtuse angles (109°28′) of the rhombs meet in the trihedral angles of the solid figure. The whole figure may be regarded as composed of six square pyramids, each with its alternate slant faces perpendicular to one another, placed on six squares forming the sides of a cube. The long diagonal of each rhombic face thus made up of two sides of pyramids conterminous in the short diagonal, is $\sqrt{2}$ times the short diagonal.

**I see it inadvertently stated by Plateau that all the twelve films are "légèrement courbées".

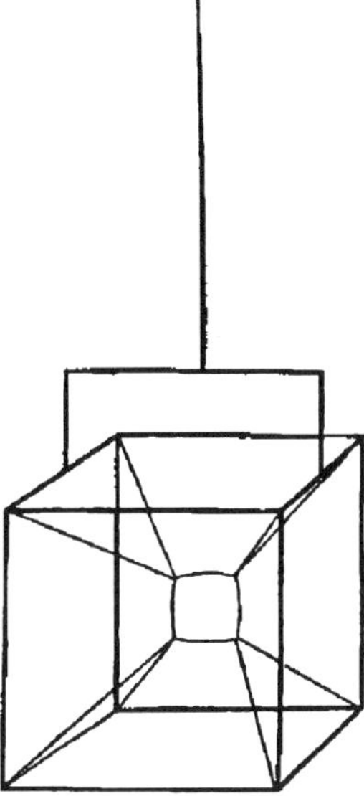

Fig. 1.

6. Suppose now space to be filled with equal and similar ideal rhombic dodecahedrons. Draw the short diagonal of every rhombic face, and fix a real wire (infinitely thin and perfectly stiff) along each. This fills space with Plateau cubic frames. Fix now, ideally, a very small rigid globe at each of the points of space occupied by tetrahedral angles of the dodecahedrons, and let the faces of the dodecahedrons be realized by soap-films. They will be in *stable* equilibrium, because of the little fixed globes; and the equilibrium would be stable without the rigid diagonals which we require only to help the imagination in what follows. Let an exceedingly small force, like gravity*, act on all the films everywhere perpendicularly to one set of parallel faces of the cubes. If this force is small enough it will not tear away the films from the globes; it will only produce a very slight bending from the plane rhombic shape of each film. Now annul the little globes. The films will instantly jump (each set of twelve which meet in a point) into the Plateau configuration (Fig. 1), with the little curve-edged square in the plane perpendicular to the determining force, which may now be annulled, as we no longer require it. The rigid edges of the cubes may also be now annulled, as we have done with them also; because each is (as we see by

*To do for every point of meeting of twelve films what is done by blowing in the experiment of Section 5.

symmetry) pulled with equal forces in opposite directions, and therefore is not required for the equilibrium, and it is clear that the equilibrium is stable without them*.

7. We have now space divided into equal and similar tetrakaidecahedral cells by the soap-film; each bounded by:

(1) Two small plane quadrilaterals parallel to one another;

(2) Four large plane quadrilaterals in planes perpendicular to the diagonals of the small ones;

(3) Eight non-plane hexagons, each with two edges common with the small quadrilaterals, and four edges common with the large quadrilaterals.

The films seen in the Plateau cube show one complete small quadrilateral, four halves of four of the large quadrilaterals, and eight halves of eight of the hexagons, belonging to six contiguous cells; all mathematically correct in every part (supposing the film and the cube-frame to be infinitely thin). Thus we see all the elements required for an exact construction of the complete tetrakaidecahedron. By making a clay model of what we actually see, we have only to complete a symmetrical figure by symmetrically completing each half-quadrilateral and each half-hexagon, and putting the twelve properly together, with the complete small quadrilateral, and another like it as the far side of the 14-faced figure. We thus have a correct solid model.

8. Consider now a cubic portion of space containing a large number of such cells, and of course a large but a comparatively small number of partial cells next the boundary. Wherever the boundary is cut by film, fix stiff wire; and remove all the film from outside, leaving the cubic space divided stably into cells by films held out against their tension by the wire network thus fixed in the faces of the cube. If the cube is chosen with its six faces parallel to the three pairs of quadrilateral films, it is clear that the resultant of the whole pull of film on each face will be perpendicular to the face, and that the resultant pulls on the two pairs of faces parallel to pairs of the greater quadrilaterals are equal to one another and less than the resultant pull on the pair of faces parallel to the smaller quadrilaterals. Let now the last-mentioned pair of faces of the cube be allowed to yield to the pull inwards,

---

*The corresponding two-dimensional problem is much more easily imagined; and may probably be realized by aid of moderately simple appliances.

Between a level surface of soap-solution and a horizontal plate of glass fixed at a centimetre or two above it, imagine vertical film-partitions to be placed along the sides of the squares indicated in the drawing (Fig. 2): these will rest in stable equilibrium if thick enough wires are fixed vertically through the corners of the squares. Now draw away these wires downwards into the liquid: the equilibrium in the square formation becomes unstable, and the films instantly run into the hexagonal formation shown in the diagram; provided the square of glass is provided with vertical walls (for which slips of wood are convenient), as shown in plan by the black border of the diagram. These walls are necessary to maintain the inequality of pull in different directions which the inequality of the sides of the hexagons implies. By inspection of the diagram we see that the pull is T/a per unit area on either of the pair of vertical walls which are perpendicular to the short sides of the hexagons; and on either of the other pair of walls $2\cos 30^\circ \times T/a$; where T denotes the pull of the film per unit breadth, and a the side of a square in the original formation. Hence the ratio of the pulls per unit of area in the two principal directions is as 1 to 1.732.

Fig. 2.

while the other two pairs are dragged outwards against the pulls on them, so as to keep the enclosed volume unchanged; and let the wirework fixture on the faces be properly altered, shrunk on two pairs of faces, and extended on the other pair of faces, of the cube, which now becomes a square cage with distance between floor and ceiling less than the side of the square. Let the exact configuration of the wire everywhere be always so adjusted that the cells throughout the interior remain, in their altered configuration, equal and similar to one another. We may thus diminish, and if we please annul, the difference of pull per unit area on the three pairs of sides of the cage. The respective shrinkage-ratio and extension-ratio, to exactly equalize the pulls per unit area on the three principal planes (and therefore on all planes), are $2^{-1/3}$, $2^{1/6}$, $2^{1/6}$, as is easily seen from what follows.

9. While the equalization of pulls in the three principal directions is thus produced, work is done by the film on the moving wire-work of the cage, and the total area of film is diminished by an amount equal to W/T, if W denote the whole work done, and T the pull of the film per unit breadth. The change of shape of the cage being supposed to be performed infinitely slowly, so that the film is always in equilibrium throughout, the total area is at each instant a minimum, subject to the conditions:

(1) That the volume of each cell is the given amount;
(2) That every part of the wire has area edged by it; and
(3) That no portion of area has any free edge.

10. Consider now the figure of the cell (still of course a tetrakaidecahedron) when the pulls in the three principal directions are equalized, as described in Section 8. It must be perfectly isotropic in respect to these three directions. Hence the pair of small quadrilaterals must have become enlarged to equality with the two pairs of large ones, which must have become smaller in the deformational process described in Section 8. Of each hexagon three edges coincide with edges of quadrilateral faces of one cell; and each of the three others coincides with edges of three of the quadrilaterals of one of the contiguous cells. Hence the 36 edges of the isotropic tetrakaidecahedron are equal and similar plane arcs; each of course symmetrical about its middle point. Every angle of meeting of edges is essentially 109°28′ (to make trihedral angles between tangent planes of the films meeting at 120°). Symmetry shows that the quadrilaterals are still plane figures; and therefore, as each angle of each of them is 109°28′, the change of direction from end to end of each arc-edge is 19°28′. Hence each would be simply a circular arc of 19°28′, if its curvature were equal throughout; and it seems from the complete mathematical investigation of Sections 16, 17, 18 below, that it is nearly so, but not exactly so even to a first approximation.

Of the three films which meet in each edge, in three adjacent cells, one is quadrilateral and two are hexagonal.

11. By symmetry we see that there are three straight lines in each (non-plane) hexagonal film, being its three long diagonals; and that these three lines, and therefore the six angular points of the hexagon, are all in one plane. The arcs composing its edges are not in this plane, but in planes making, as we shall see (Section 12), angles of 54°44′ with it. For three edges of each hexagon, the planes of the arcs bisect the angle of 109°28′ between the planes of the six corners of contiguous hexagons; and for the other three edges are inclined on the outside of its plane of corners, at angles equal to the supplements of the angles of 125°16′ between its plane of corners and the planes of contiguous quadrilaterals.

12. The planes of corners of the eight hexagons constitute the faces of an octahedron which we see, by symmetry, must be a regular octahedron (eight equilateral triangles in planes inclined 109°28′ at every common edge). Hence these planes, and the planes of the six quadrilaterals, constitute a plane-faced tetrakaidecahedron obtained by truncating the six corners* of a regular octahedron each to such a depth as to reduce its eight original (equilateral triangular) faces to equilateral equiangular hexagons. An orthogonal projec-

*This figure (but with probably indefinite extents of the truncation) is given in books on mineralogy as representing a natural crystal of red oxide of copper.

tion of this figure is shown in Fig. 3. It is to be remarked that space can be filled with such figures. For brevity we shall call it a plane-faced isotropic tetrakaidecahedron.

13. Given a model of the plane-faced isotropic tetrakaidecahedron, it is easy to construct approximately a model of the *minimal tetrakaidecahedron*, thus: —Place on each of the six square faces a thin plane disk having the proper curved arcs of 19°28′ for its edges. Draw the three long diagonals of each hexagonal face. Fill up by little pieces of wood, properly cut, the three sectors of 60° from the centre to the overhanging edges of the adjacent quadrilaterals. Hollow out symmetrically the other three sectors, and the thing is done. The result is shown in orthogonal projection, so far as the edges are concerned, in Fig. 4; and as the orthogonal projections are equal and similar on three planes at right angles to one another, this diagram suffices to allow a perspective drawing from any point of view to be made by "descriptive geometry".

14. No shading could show satisfactorily the delicate curvature of the hexagonal faces, though it may be fairly well seen on the solid model made as described in Section 12. But it is shown beautifully, and illustrated in great perfection, by making a skeleton model of 36 wire arcs for the 36 edges of the complete figure, and dipping it in soap solution to fill the faces with film, which is easily done for all the faces but one. The

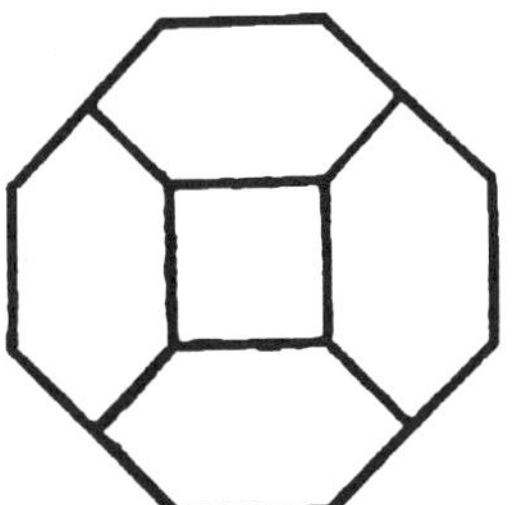

Fig. 3.

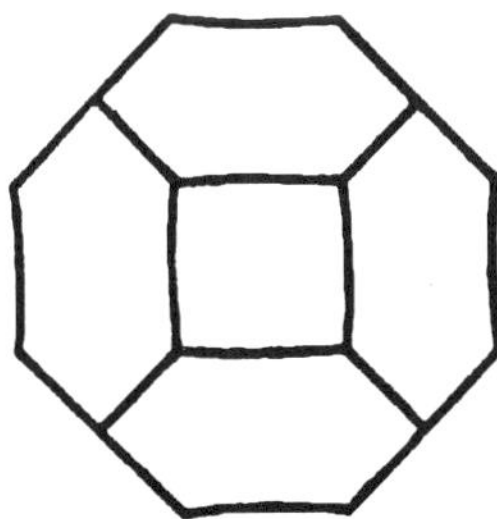

Fig. 4.

curvature of the hexagonal film on the two sides of the plane of its six long diagonals is beautifully shown by reflected light. I have made these 36 arcs by cutting two circles, 6 inches diameter, of stiff wire, each into 18 parts of 20° (near enough to 19°28′). It is easy to put them together in proper positions and solder the corners, by aid of simple devices for holding the ends of the three arcs together in proper positions during the soldering. The circular curvature of the arcs is not mathematically correct, but the error due to it is, no doubt, hardly perceptible to the eye.

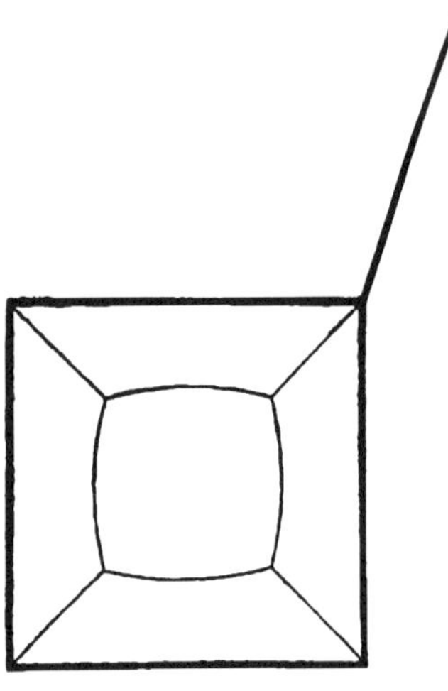

Fig. 5.

Fig. 6.

15. But the true form of the curved edges of the quadrilateral plane films, and of the non-plane surfaces of the hexagonal films, may be shown with mathematical exactness by taking, instead of Plateau's skeleton cube, a skeleton square cage with four parallel edges each 4 centimetres long: and the other eight, constituting the edges of two squares each $\sqrt{2}$ times as long, or 5.66 centim. Dipped in soap-solution and taken out it always unambiguously gives the central quadrilateral in the plane perpendicular to the four short edges. It shows with mathematical accuracy (if we suppose the wire edges infinitely thin) a complete quadrilateral, four half-quadrilaterals, and four half-hexagons of the minimal tetrakaidecahedron. The two principal views are represented in Figs. 5 and 6.

16. The mathematical problem of calculating the forms of the plane arc-edges, and of the curved surface of the hexagonal faces, is easily carried out to any degree of approximation that may be desired; though it would be very laborious, and not worth the trouble, to do so further than a first approximation, as given in Section 17 below. But first let us state the rigorous mathematical problem; which by symmetry becomes narrowed to the consideration of a 60° sector BCB′ of our non-plane hexagon, bounded by straight lines CB, CB′ and a slightly curved edge BEB′, in a plane, Q, through BB′, inclined to the plane BCB′ at an angle of $\tan^{-1}\sqrt{2}$, or 54°44′. The plane of the curved edge I call Q, because it is the plane of the contiguous quadrilateral. The mathematical problem to be solved is *to find the surface of zero curvature edged by* BCB′ *and cutting at* 120° *the plane*

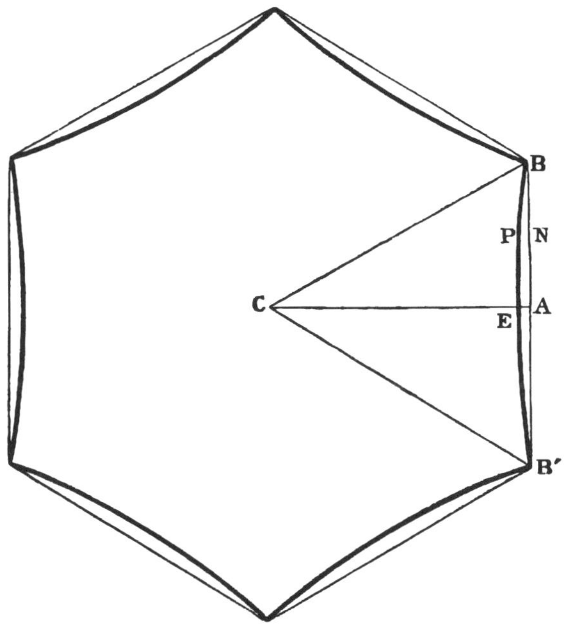

Fig. 7.

Q *all along the intersectional curve* (Fig. 7). It is obvious that this problem is determinate and has only one solution. Taking CA for axis of x; and z perpendicular to the plane BCB′: and regarding z as a function of x, y, to be determined for finding the form of the surface, we have, as the analytical expression of the conditions

$$\frac{d^2z}{dx^2}\left(1+\frac{dz^2}{dy^2}\right) - 2\frac{dz}{dx}\frac{dz}{dy}\frac{d^2z}{dxdy} + \frac{d^2z}{dy^2}\left(1+\frac{dz^2}{dx^2}\right) = 0; \tag{1}$$

and

$$\left(1+\frac{dz^2}{dx^2}+\frac{dz^2}{dy^2}\right)^{-1/2}\left(\sqrt{\frac{1}{3}}-\frac{dz}{dx}\sqrt{\frac{2}{3}}\right)=\frac{1}{2} \quad \text{when} \quad z=(a-x)\sqrt{2}. \tag{2}$$

17. The required surface deviates so little from the plane BCB′ that we get a good approximation to its shape by neglecting dz^2/dx^2, $dz/dx{\cdot}dz/dy$, and dz^2/dy^2, in (1) and (2), which thus become

$$\nabla^2 z = 0, \tag{3}$$

and

$$\frac{dz}{dx}=\frac{\sqrt{2}}{2}-\sqrt{\frac{3}{8}}=.94735, \quad \text{when} \quad x=a-z/\sqrt{2}, \tag{4}$$

∇^2 denoting $(d/dx)^2 + (d/dy)^2$. The general solution of (3), in polar coordinates (r, ϕ) for the plane (x, y), is

$$\sum (A\cos m\phi + B\sin m\phi) r^m, \tag{5}$$

where A, B, and m are arbitrary constants. The symmetry of our problem requires $B = 0$, and $m = 3{\cdot}(2i + 1)$, where i is any integer. We shall not take more than two terms. It seems not probable that advantage could be gained by taking more than two, unless we also fall back on the rigorous equations (1) and (2), keeping dz^2/dx^2 & c. in the account, which would require each coefficient A to be not rigorously constant but a function of r. At all events we satisfy ourselves with the approximation yielded by two terms, and assume

$$z = Ar^3\cos 3\phi + A'r^9\cos 9\phi; \tag{6}$$

with two coefficients A, A' to be determined so as to satisfy (4) for two points of the curved edge, which, for simplicity, we shall take as its middle, E ($\phi = 0$); and end, B ($\phi = 30°$). Now remark that, as z is small, even at E, where it is greatest, we have, in (4), $x \doteqdot a$ or $r \doteqdot a\sec\phi$. Thus, and substituting for dz/dx its expression in polar (r, ϕ) coordinates, which is

$$\frac{dz}{dx} = \frac{dz}{dr}\cos\phi - \frac{dz}{rd\phi}\sin\phi, \tag{7}$$

we find, from (4) with (6),

$$(\text{by case } \phi = 0) \qquad A + 3a^6 A' = .031578a^{-2}, \tag{8}$$

$$(\text{and by case } \phi = 30°) \quad A - \frac{64}{9}a^6 A' = .031578 \cdot \frac{3}{2} \cdot a^{-2}; \tag{9}$$

whence

$$A' = -\frac{1}{2} \cdot \frac{9}{91} \times .031578 \cdot a^{-8} = -9 \times .0001735 \cdot a^{-8} = -.001561 \cdot a^{-8},$$

$$A = \frac{1}{2}\left(3 - \frac{64}{91}\right) \times .031578 \cdot a^{-2} = 209 \times .0001735 \cdot a^{-2} = .036261 \cdot a^{-2};$$

and for required equation of the surface we have (taking $a = 1$ for brevity)

$$\left.\begin{aligned} z &= .03626 \cdot r^3 \cos 3\phi - .001561 r^9 \cos 9\phi \\ &= .03626 \cdot r^3 \left(\cos 3\phi - .043 \cdot r^6 \cos 9\phi\right) \end{aligned}\right\}. \tag{10}$$

18. To find the equation of the curved edge BEB$'$, take, as in (4),

$$x = 1 - z\sqrt{2} = 1 - \xi, \qquad \text{where } \xi \text{ denotes } z\sqrt{2}. \tag{11}$$

Substituting in this, for z, its value by (10), with for r its approximate value $\sec\phi$, we find

$$\xi = \frac{1}{\sqrt{2}}\left(.03626 \sec^3\phi \cos 3\phi - .001561 \sec^9\phi \cos 9\phi\right) \tag{12}$$

as the equation of the orthogonal projection of the edge, on the plane BCB′, with

$$\mathrm{AN} = y = \tan\phi; \quad \text{and} \quad \mathrm{NP} = \xi. \tag{13}$$

The diagram was drawn to represent this projection roughly, as a circular arc, the projection on BCB′ of the circular arc of 20° in the plane Q, which, before making the mathematical investigation, I had taken as the form of the arc-edges of the plane quadrilaterals. This would give 1/35 of CA, for the sagitta, AE; which we now see is somewhat too great. Equation (12), with $y = 0$, gives for the sagitta

$$\mathrm{AE} = .0245 \times \mathrm{CA}, \tag{14}$$

or, say, 1/41 of CA. The curvature of the projection at any point is to be found by expressing $\sec^3\phi\cos 3\phi$ and $\sec^9\phi\cos 9\phi$ in terms of $y = \tan\phi$ and taking d^2/dy^2 of the result.

By taking $\sqrt{2/3}$ instead of $\sqrt{1/2}$ in (12), we have the equation of the arc itself in the plane Q.

19. To judge of the accuracy of our approximation, let us find the greatest inclination of the surface to the plane BCB′. For the tangent of the inclination at (r, ϕ) we have

$$\left(\frac{dz^2}{dr^2} + \frac{dz^2}{r^2 d\phi^2}\right)^{1/2} = .1088 \cdot r^2\left(1 - 2 \times .129 \cdot r^6 \sec 6\phi + .129^2 r^{12}\right)^{1/2}. \tag{15}$$

The greatest values of this will be found at the curved bounding edge, for which $r \doteqdot \sec\phi$. Thus we find

$$\left.\begin{array}{l} \left(1 + \dfrac{dz^2}{dr^2} + \dfrac{1}{r^2}\dfrac{dz^2}{d\phi^2}\right)^{1/2} \\ \left\{\begin{array}{l} = .0948, \text{ and therefore inclination } = 5°\,25' \text{ at E} \\ = .1892, \text{ and therefore inclination } = 10°\,43' \text{ at B} \end{array}\right\} \end{array}\right\}. \tag{16}$$

Hence we see that the inaccuracy due to neglecting the square of the tangent of the inclination in the mathematical work cannot be large. The exact value of the inclination at E is $\tan^{-1}(-\sqrt{2}) - 120°$, or 5°16′, which is less by 9′ than its value by (16).

Seifenblasenrhetorik, or the Sound of the Last Trumpet?

N. RIVIER

Laboratoire de Physique Théorique, Université Louis Pasteur, F 67084 Strasbourg

Keywords: Foam, Packing, Honeycomb, Tiling, Frustration

"The geometry of the 14-hedron, displayed to perfection by our soap films in the twinkling of an eye ... ." This is how D'Arcy THOMPSON (1917, 1942) describes the result of Lord Kelvin's "short but very beautiful paper" (THOMSON, 1887) with which this volume of Forma is concerned. Sir D'Arcy does not add "... at the last trumpet, ... and the truth shall be revealed, incorruptible", but this is very much in the mind of the reader, carried away by his prose or by the gusto of Handel's "The trumpet shall sound." Was the truth revealed, or was it only a lot of hot air (Seifenblasenrhetorik)?

Can a froth be regular? What are the shapes of its bubbles? If the froth is random, is there a most probable shape? In two dimensions, all the answers are realized by the honeycomb. In three dimensions, in total contrast, froths are almost all random, except for the theoretical solution of Kelvin (THOMSON, 1887), for the two-cell partition of WEAIRE and PHELAN (1994) and for its nearest relatives, the tetrahedrally close-packed structures (tcp) (CHARVOLIN and SADOC, 1988; RIVIER, 1994; RIVIER and ASTE, 1996). Three-dimensional froths are also geometrically frustrated, in that closest packing is not compatible with a regular structure (Note 1).

Our story begins with the Reverend Stephen Hales, a distinguished scientist (his face adorns the National Portrait Gallery), who published in 1727 an experiment on sap flow (HALES, 1727). He determined the Voronoi froth associated with peas, which, after taking up water, deformed into "pretty regular Dodecahedrons". As ZALLEN (1983) and COXETER (1961) emphasize, the reference to dodecahedra is a qualitative observation of the predominance of pentagonal faces, not a geometrical theorem. However, the dodecahedron became the Received Knowledge for the shape of cells in a regular cellular froth governed by surface tension, which remained unchallenged for a century and a half, until Kelvin (THOMSON, 1887). Even worse, "regular" had been changed into "rhombic" (apparently by BUFFON (1753). See THOMPSON (1917, 1942), p. 552; BONNER (1961), p. 122. This is not twinkling of an eye, but wishful thinking: Even if the garden peas had been carefully packed in a face-centered cubic array, they would only swell into rhombic dodecahedra if they had all exactly the same size and spherical shape, which clearly is beyond the ability of a gardener as intelligent as Mendel's.) (Note 2).

After Kelvin's paper and the first edition of *On Growth and Form*, it was the botanists who became interested in the structure of froths. Indeed, undifferentiated tissues like the apple and the wing of a bird adorning the frontispiece of Dormer's textbook on geometry for biologists (DORMER, 1980), are undistinguishable, not only from each other, but from a soap froth. The bee's honeycomb is a hexagonal partition of two-dimensional space, but it is built by the bees piling on top of each other, filling the interspace with building wax and letting it dry (although, if left by themselves, bees will make larger cells on the periphery to accommodate drones. They are separated from the workers cells by beautiful arrays of dislocations, forming grain boundaries (Fig. 1)). What about three-dimensional biological tissues?

The main contributions were those of LEWIS (1928), MATZKE (1946) and MARVIN (1939). One of the questions was to find out whether a biological tissue was shaped by surface tension (like a soap froth) or by packing (like compressed lead-shot, an experiment that the botanist MARVIN (1939) did well before the physicist BERNAL's (1959) classic realization of the structure of an atomic liquid). In fact, it matters little and the structure is moulded above all by disorder (maximum entropy) and by steric or topological constraints. In retrospect, the first indication (1928) of this universality of topological

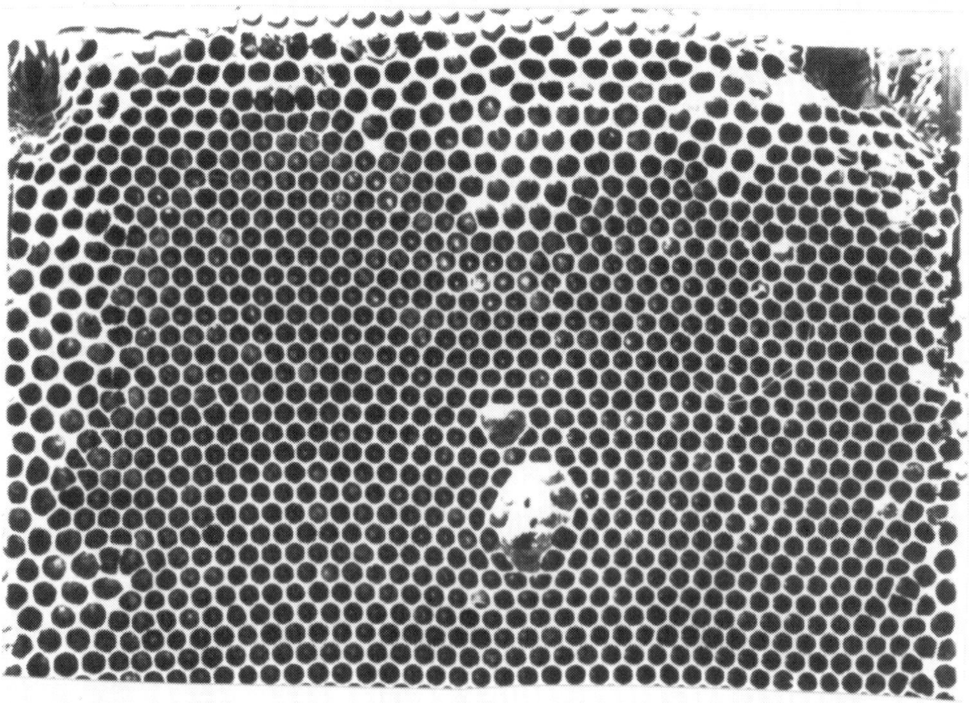

Fig. 1. Part of a bee's honeycomb. The larger cells on the periphery accommodate drones. They are separated from the workers' cells by dislocations (Courtesy: J.-C. Garaud).

froths was the famous law of LEWIS (1928) discovered on the epithelia of two cucumbers, one retina and one human amnion. Much later came the metallurgists SMITH (1952) and ABOAV (1970).

Marvin, Matzke and Lewis noticed the near-absence of Kelvin's tetrakaidecahedra, the abundance of pentagonal facets and the disorder (Notes 3, 4). This culminated in Matzke's celebrated after-dinner speech "In the Twinkling of an Eye" (MATZKE, 1950), which knocked down order, Kelvin and D'Arcy Thomson for another forty years. Indeed, in the third, abridged edition (1961) of *On Growth and Form* (BONNER, 1961), its Editor, J. T. Bonner, himself a distinguished specialist of morphogenesis, has added several comments, based mostly on Matzke's speech. "All that remains then for Kelvin's ideal 14-hedra is the frictionless world of soap-bubbles. But here again *expectations do not fit the facts* .... There is a danger in making hypotheses that can be tested, but even with the new information before us the problem has not lost its interest .... In all these instances the "cells" are irregular, (which) may indicate that ... there are a set of forces acting that will automatically frustrate any perfect ... configuration." (This was well before the geometrical frustration of the problem of 3-dimensional packing was identified and resolved by Sadoc, Kléman and Charvolin (KLÉMAN and SADOC, 1979; CHARVOLIN and SADOC, 1988).)

It took some time for somebody to face the danger in "making testable hypotheses". This somebody was in fact the Editor of this volume, who first suggested that Kelvin's polyhedra could be seen at least in froths confined in cylindrical tubes, then found (with Pittet, Hutzler and Pardal (WEAIRE *et al.*, 1993)) that the confined foam remained structurally unchanged through a transition by drainage from polyhedral to spherical bubbles, and finally, with Phelan (WEAIRE and PHELAN, 1994), that the A15 structure had in fact a lower energy (and many close relatives to visit, thereby increasing its entropy at little cost in energy, and ending up disordered) than the splendid but isolated and still elusive construction of Kelvin's (Note 5).

Notes

1. A froth is a partition of D-dimensional space by convex cells. Randomness, or absence of adjustment imposes that each vertex has minimal number of incident edges, faces and cells. (The USA do not constitute a political froth, because the Four Corners Boundary between Utah, Arizona, New Mexico and Colorado has obviously been adjusted.) The boundary of each D-cell of a froth is itself a froth in elliptic (positively-curved) (D − 1)-dimensional space. The surface of a polyhedral bubble of a three-dimensional froth is a curved two-dimensional froth with polygonal cells, which are the interfaces beween the bubbles of the original froths. A froth is therefore a graded set of topological froths with minimal incidence.

2. Rhombic dodecahedra are the Voronoi cells of face-centered cubic packing (like the oranges at the grocer). Because each cell has some 4-coordinated vertices, it is not a topologically stable polyhedron, and cannot be a suitable cell of any random packing. The packing is indeed "adjusted". Nobody who have seen compressed peas, lead shot or soap froth can possibly suggest the rhombic dodecahedron as a typical cell, unless they

were thinking of packing cannon balls (an activity more likely to make one the "Ruler of the Queen's Navy" than playing with soap bubbles).

3. Matzke's experiments seem to have been made carefully, and with an open mind. I still have to see in three-dimensional froths more than a few % of Kelvin's 14-hedra, even deformed. (Surface cells are excluded. For the tube geometry of WEAIRE *et al.* (1993), there should be at least three shells of inner cells for the froth to be deemed three-dimensional.) Clathrate cells, like the Goldberg 14-hedron made of 2 hexagons and 12 pentagons or the other tcp cells (pentagonal dodecahedron, a 15-hedron and a 16-hedron, all with 12 pentagons and no, 3 and 4 hexagons respectively) can indeed be found in foams.

4. Kelvin's 14-hedron space-filler is a truncated cube or truncated octahedon, with 6 square faces and 8 hexagonal faces. The Weaire-Phelan A15 structure (WEAIRE and PHELAN, 1994) is a space-filling combination of one pentagonal dodecahedron to 3 Goldberg 14-hedra with 12 pentagonal and two hexagonal faces.

5. These tetrahedrally close-packed configurations have all very low energy. They are similar and connected to each other by elementary topological transformations. A disordered mixture of these local configurations costs very little energy while increasing the entropy and decreasing the free energy of the structure. The Kelvin froth stands alone, in splendid isolation from the tcp range, with a low energy whereas its local elementary excitatons are very costly in energy (RIVIER, 1994; RIVIER and ASTE, 1996).

This work is supported by the EC Network: Physics of Foams, CHRXCT 940542.

REFERENCES

ABOAV, D. A. (1970) *Metallogr.*, **3**, 383.
BERNAL, J. D. (1959) *Nature*, **183**, 141.
BONNER, J. T. (ed.) (1961) D'Arcy Thomson, *On Growth and Form*, 3rd edition, Cambridge Univ. Press, ch.4.
BUFFON, G. L. L. (1753) *Histoire Naturelle*, Paris.
CHARVOLIN, J. and SADOC, J. F. (1988) *J. Physique (Paris)*, **49**, 521.
COXETER, H. S. M. (1961) *Introduction to Geometry*, Wiley, New York.
DORMER, K. J. (1980) *Fundamental Tissue Geometry for Biologists*, Cambridge Univ. Press.
HALES, S. (1727) *Vegetable Staticks*, Innys and Woodward, London.
KLÉMAN, M. and SADOC, J. F. (1979) *J. Physique Lett.*, **40**, 569.
LEWIS, F. T. (1928) *Anat. Records*, **38**, 341; (1943) *Am. J. Botany*, **30**, 74.
MARVIN, J. W. (1939) *Amer. J. Bot.*, **26**, 280.
MATZKE, E. B. (1946) *Am. J. Botany*, **33**, 58.
MATZKE, E. B. (1950) *Bull. Torrey Bot. Club*, **27**, 222.
RIVIER, N. (1994) *Phil. Mag. Letters*, **69**, 297.
RIVIER, N. and ASTE, T. (1996) *Forma*, **11**, **61–9**.
SMITH, C. S. (1952) in *Metal Interfaces*, Amer. Soc. Metals, Cleveland, 65 pp.
THOMPSON, D'A. W. (1917), (1942) *On Growth and Form*, Cambridge Univ. Press, ch.7.
THOMSON, W. (Lord Kelvin) (1887) *Phil. Mag.*, **25**, 503.
WEAIRE, D. and PHELAN, R. (1994) *Phil. Mag. Letters*, **69**, 107.
WEAIRE, D., PITTET, N., HUTZLER, S. and PARDAL, R. (1993) *Phys. Rev. Letters*, **71**, 2670.
ZALLEN, R. (1983) *The Physics of Amorphous Solids*, Wiley, New York; (1979) in *Fluctuation Phenomena* (eds. E. W. Montroll and J. L. Lebowitz), North Holland, Amsterdam, pp. 207.

Soap Bubble Clusters

F. ALMGREN[1] and J. E. TAYLOR[2]*

[1]*Department of Mathematics, Princeton University,*
Fine Hall, Washington Road, Princeton, NJ 08544, U.S.A.
[2]*Department of Mathematics, Rutgers University,*
Hill Center, Busch Campus, Piscataway, NJ 08855, U.S.A.

Keywords: Geometric Measure Theory, Minimal Surface, Soap Bubbles

1. A Glorious History

"Surface tension phenomena have attracted the interest of many mathematicians and physicists. According to MAXWELL, Leonardo da Vinci is credited with the discovery of capillary action (which is another manifestation of surface energy effects, involving in this case the media air, water or some other liquid, and glass), Isaac Newton was the first to produce the idea of molecular forces affecting the cohesion of liquids, and Laplace, Gauss, and Poisson all worked on the theory of how the surface energies arose (from thermodynamics) and the constraints that minimizing surface energy put on the resulting forms of the surfaces (from mathematics). The culmination of the thermodynamic approach was the work of GIBBS (1928), whereas the major experimental study of the shapes of surfaces governed by surface tension, in particular soap films, was that of PLATEAU (1873). ... Although in the past 100 years, several mathematical models for area minimizing surfaces—all called Plateau's problem—have been constructed and the regularity and singularity structure of the resulting surfaces have been more or less extensively studied (in particular, see DOUGLAS (1931), RADO (1933), OSSERMAN (1970)), none of these models allowed the general kind of surfaces which arise in real soap films. Only with the introduction of varifolds by ALMGREN (1965) has the development of a model for studying such surfaces been possible" (from TAYLOR, 1976).

For a general discussion and summary of the thermodynamics of multicomponent, multiphase systems in which interface energy is a significant component of the relevant free energy, see CAHN and TAYLOR (1996). For a discussion in non-technical terms of the ideas of the mathematics outlined in Sections 2–5 below, see ALMGREN and TAYLOR (1976).

*Presently visiting the Institute for Advanced Study, Princeton, NJ 08540, U.S.A.

2. A Mathematical Soap Bubble Problem

Among all partitionings of space into bounded regions $A_1, A_2, A_3, \ldots, A_N$ of prescribed volumes together with an outside region A_0, is there one which has least total interface area

$$\frac{1}{2}\left(|\partial A_0| + |\partial A_1| + |\partial A_2| + |\partial A_3| + \cdots + |\partial A_N|\right)?$$

If so, what is its structure? (The 1/2 is because each interface is being counted twice.)

This is one possible formulation of questions about the global existence and local structure of mathematical configurations modeled on the soap bubble geometries of nature. As the problem is stated, it does not include partitionings of all of space into regions of equal volumes having least average interface area as in Kelvin's problem. It seems to the authors that there are at least several different ways in which such problems with infinitely many regions might be formulated mathematically. Although it is plausible to them that solutions exist to such problems, they know no general theorems which guarantee this without additional work. The basic obstacle is that the diameter of regions $A_{i_0}(j)$ in a minimizing sequence of such partitions might grow without bound as $j \to \infty$. This might happen even if reasonable assumptions of periodicity are imposed on minimizing sequences. We therefore limit ourselves to the formulation given above for existence results.

The example of three colinear bubbles (where one pair has no interface) shows that soap bubble clusters need not be area minimizing among all possible configurations having the same distribution of volumes (since three mutually interfacing bubbles have less area), but only minimizing among those nearby configurations to which they could readily change. The regularity and singularity structure theorems below applies to clusters which are area minimizing among nearby configurations as well as to globally minimizing clusters.

One of the reasons this is a hard problem is that there is no *a priori* upper bound to the possible geometric or topological complexity of the configurations of the solutions one seeks. The formulation of the problem itself guarantees the existence of disjointed regions $A_i(j)$ for $i = 0, 1, \ldots, N$ and $j = 1, 2, 3, \ldots$ such that the sum of the interface areas

$$\frac{1}{2}\left(|\partial A_0(j)| + |\partial A_1(j)| + |\partial A_2(j)| + |\partial A_3(j)| + \cdots + |\partial A_N(j)|\right)$$

approaches the infimum of possible values as $j \to \infty$. Even in the case in which $N = 1$ there are minimizing sequences $\{A_1(j)\}_j$ for which every point in space is a limit as $j \to \infty$ of points $p_j \in \partial A_1(j)$; see for example Fig. 1.3.4 of MORGAN (1995). The possibility of such behavior and the absence of other complexity bounds is the reason that general existence and regularity theorems were proved only after the development of geometric measure theory; see, for example, ALMGREN (1992).

3. Existence of Soap Bubbles

The existence theorem for mathematical soap bubble clusters is based on the following fundamental theorem.

Suppose $A(1), A(2), A(3), \ldots$ are regions satisfying the following conditions:

(a) The union $\cup j\, A(j)$ of all the regions lies in a bounded region of space.

(b) The areas $|\partial A(j)|$ of the boundaries $\partial A(j)$ of the regions are uniformly bounded. Under these assumptions there will exist a subsequence $j(1), j(2), j(3), \ldots$ of $1, 2, 3, \ldots$ and a limit region A such that

(1) the volume differences $|A \sim A(j(i))| + |A(j(i)) \sim A|$ converge to zero as $i \to \infty$ (this implies that $|A| = \lim_{i\to\infty}|A(j(i))|$) and

(2) the area $|\partial A|$ of the limit region is not more than the lower limit of the areas $|\partial A(j(i))|$ as $i \to \infty$.

This is a special case of the compactness theorem for integral currents which can be proved (with some work) by applying Ascoli's theorem to multi-function representations of integral currents; see ALMGREN (1986). In order to apply the compactness theorem to the problem at hand, one has to modify a minimizing sequence $\{A_i(j)\}_{i,j}$ to keep the bubble clusters bounded in space while preserving their volumes; the construction of Chapter VI of ALMGREN (1976) accomplishes this. The formal mathematical existence of soap bubble clusters follows by applying the compactness theorem to subsequences of subsequences of the indexing sets to obtain simultaneous limits $A_1, A_2, \ldots, A_N$ together with the outside A_0.

4. The Almost Everywhere Regularity of Mathematical Soap Films

The type of interfaces ∂A_i which the compactness theorem above produces are so called *rectifiable* sets. This means, in particular, that the set can be changed on a set of small measure to become a smooth surface. The complication is that the complexity of the surface may increase without bound as the area on which the change is permitted becomes smaller and smaller. In order to establish that the solutions to our soap bubble problem have more structure, one approach is to invoke the theory of (γ, δ) *restricted sets* and the theory of $(\mathbf{M}, \varepsilon, \delta)$ *minimal sets* as set forth in chapters II and III of ALMGREN (1976).

4.1. (γ, δ) restricted sets

A compact subset S of space is called (γ, δ) restricted provided no Lipschitz deformation of space, which moves only a region W of diameter at most δ, can decrease the area of $S \cap W$ by more that a factor of γ. The reason for introducing this notion is to isolate certain parts of the geometric analysis in a way that the actual dependence upon hypotheses becomes more apparent. Examples of (γ, δ) restricted sets include

- The boundary ∂A_i of a single region in a soap bubble cluster.
- The union $\cup_i \partial A_i$ of the boundaries of all the regions in a soap bubble cluster.
- Any compact continuously differentiable submanifold without boundary.

The basic facts about such compact sets S include

(1) S is rectifiable.

(2) The upper and lower density ratios are bounded for radii comparable to δ.

It is the fact (2) which enables one to separate almost everywhere the union of the ∂A_i's into pieces of surface which separate exactly two of the ∂A_i's, i.e. the solution surface does not get impossibly tangled up with itself.

4.2. (M, ε, δ) *minimal sets*

A surface S is called (**M**, ε, δ) minimal provided no Lipschitz deformation of space, which moves only a region W of diameter r, not exceeding $\delta > 0$, can decrease the area of $S \cap W$ by more that a factor of $(1 + \varepsilon(r))$. The reason for introducing this notion is to incorporate regularity theorems for solutions to a wide class of problems into a single mathematical framework. Examples of (**M**, ε, δ) minimal sets include the following (in each case $\varepsilon(r) = Cr$ for a suitable constant C):

- The union $\cup_i \partial A_i$ of the boundaries of all the regions in a soap bubble cluster.
- Any surface which minimizes area subject to the constraint of avoiding an open set having smooth boundary.
- Any surface which minimizes the sum of its area and the bulk integral over, say, the region enclosed of a positive continuous function.

The basic fact about such sets S is the following "almost everywhere regularity theorem" (ALMGREN, 1976):

Suppose S is (**M**, ε, δ) *minimal with* $\varepsilon(r) = Cr$ *for a suitable constant C. Then S contains a (possibly empty) compact subset* Σ *having zero two dimensional measure such that* $S \sim \Sigma$ *is a Hölder continuously differentiable submanifold.*

Several remarks are in order:

- For many problems whose solutions are (**M**, ε, δ) minimal sets, the function $\varepsilon(r)$ and the diameter δ are not determined *a priori* but depend on the particular solution.
- The deformations with respect to which one determines (**M**, ε, δ) minimality do not have to respect constraints such as, in the soap bubble problem, preserving the prescribed volumes of the separate regions.
- The notion of a given boundary B appears in the problem by requiring the deformations to be continuous everywhere and not to move B.
- For rectifiable currents which are "almost minimal" with respect to their Stokes's theorem boundaries, a simpler proof of the almost everywhere regularity is given in BOMBIERI (1982).
- To prove almost everywhere regularity of soap bubble clusters, one could avoid the (**M**, ε, δ) minimal definition and alternatively just invoke the volume adjustment construction of ALMGREN (1976) Chapter VI to infer that the soap bubble cluster $S = \cup_i \partial A_i$ is a rectifiable set whose associated integral varifold has bounded mean curvatures in the sense of ALLARD (1972), and then apply the regularity results of ALLARD (1972). But it

is precisely ($\mathbf{M}$, ε, δ) minimal sets which have soap bubble type singularities as described below.

5. The Singular Sets of Soap Films

Plateau observed experimentally that there are only two possible singularities of soap films: the *Y*, consisting of three half planes meeting along their common boundary line at 120° angles, and the *T*, which is the cone over the 1-skeleton of the regular tetrahedron. This and more was proved in TAYLOR (1976):

The singular sets of two-dimensional ($\mathbf{M}$, ε, δ) *minimal sets in* R^3 *are* $C^{1,\alpha}$ *curves with tangent cone Y (and with a* $C^{1,\alpha}$ *deformation of space taking the set to Y locally), together with isolated points with tangent cone T at which four of these singular curves meet (and with a* $C^{1,\alpha}$ *deformation of space taking the set to T locally).*

Using this $C^{1,\alpha}$ smoothness of the curve as a starting point, NITSCHE (1977) showed that the curves are in fact C^∞, and KINDERLEHRER *et al.* (1978) showed that the curves are in fact analytic. WHITE (1985) simplified several steps of the proof.

LAWLOR and MORGAN (1994) recast the question of the minimality of *T* subject to separating four regions (which was first obtained in TAYLOR (1976)) as a calibrations problem (related to max-flow-min-cut) and proved minimality directly. BRAKKE (1996) then recast it as problem in covering spaces, and proved minimality of *Y*, *T*, and many other surfaces. MORGAN and TAYLOR (1991) showed that *T* was no longer minimizing if the energy included not only surface energy but also a line energy associated to the triple-junction curves (as might happen in solid materials).

6. Anisotropy

The surface free energy of an interface is, in the simplest cases (such as models of soap bubbles), a constant times the area of the interface. In more complicated situations, such as interfaces between a crystal and its melt (e.g. ice and water), surface energy may depend on other factors such as orientation of the interface with respect to the crystal lattice or, perhaps, temperature. The dependence upon orientation is often recorded as a surface free energy density function (integrand) $\Phi(\boldsymbol{n})$ defined on all unit normal vectors $\boldsymbol{n}$ so that the surface energy of a surface S with normal vectors $\boldsymbol{n}(x)$ is

$$\Phi(S) = \int_{x \in S} \Phi(\boldsymbol{n}(x))dA.$$

The surface energy is called *isotropic* provided $\Phi(\boldsymbol{n})$ is a constant function; otherwise it is *anisotropic*. Φ is called *even* in case $\Phi(-\boldsymbol{n}) = \Phi(\boldsymbol{n})$ for each $\boldsymbol{n}$. For technical convenience (it is often easier to work in a vector space), $\Phi(v)$ is often defined for all vectors v by setting $\Phi(v) = |v|\Phi(v/|v|)$. For example, Φ is called *convex* if the extended function is convex in the usual sense for functions defined on a vector space. A convex even Φ is thus a norm. The *area integrand* is the Euclidean norm $\mathbf{M}(v) = |v|$. If a variety of different crystalline substances are present, then the interface between each pair of substances may require a separate Φ. If the substances are immiscible liquids (e.g. oil, water, mercury, air) then each $\Phi(v) = c|v|$ but with different values for c.

Just as mean curvature can be regarded as the variation of area with volume under deformations, there is a notion of weighted mean curvature, which is the variation of surface free energy with volume. For a variety of characterizations of weighted mean curvature, see TAYLOR (1992a).

6.1. Wulff shapes as anisotropic soap bubbles

Suppose one is given a continuous surface energy density function $\Phi(\boldsymbol{n})$. What is the shape of the solid of given volume having the smallest surface energy? The answer is given by Wulff's construction which produces the Wulff shape for Φ,

$$W_\Phi = \bigcap_{|\boldsymbol{n}|=1} \{x : x \cdot \boldsymbol{n} \le \Phi(\boldsymbol{n})\}.$$

As the intersection of half spaces, W_Φ is convex. In case Φ is a norm, then W_Φ is the unit ball of the dual norm Φ^* defined by setting

$$\Phi^*(v) = \sup_{|\boldsymbol{n}|=1} \frac{v \cdot \boldsymbol{n}}{\Phi(\boldsymbol{n})}.$$

If W_Φ is a polyhedron, then Φ is defined to be *crystalline*. See TAYLOR (1994) for the history of the Wulff shape and many references.

6.2. Crystal clusters

Suppose one wishes to obtain a cluster of crystal grains of prescribed volumes and of least total interface energy. Under assumptions (a) and (b) of Section 3 we conclude additionally

(3) For each Φ which is nonnegative and convex, the Φ surface energy $\Phi(\partial A)$ of the limit surface ∂A is not more than the lower limit of the surface energies $\Phi(\partial A(j(i)))$ as $i \to \infty$.

In case Φ is convex and even, then the arguments sketched for soap bubble clusters carry over and one concludes the existence of clusters of prescribed volumes and of least total Φ energy. Such clusters are (γ, δ) restricted. If additionally Φ is uniformly convex and is three times continuously differentiable, then the theory of $(\Phi, \varepsilon, \delta)$ minimal sets applies (ALMGREN, 1976) and one concludes the almost everywhere regularity of the minimizing cluster.

In the case of general polycrystals, in which each interface $\partial A_i \cap \partial A_j$ has its own surface free energy density function $\Phi_{i,j}$, the situation becomes more subtle. Under reasonable conditions, however, minimizing clusters do exist and are (γ, δ) restricted sets and, indeed, are almost everywhere regular. LAWLOR and MORGAN (1994) derived and proved a general criterion for the minimality of tangent cones at singular points (such criteria had been earlier given by Gibbs and extended by Cahn). The present state of knowledge is summarized in the thesis (CARABALLO, 1996) which treats also the evolution of such clusters as discussed below.

7. Motion

Work on soap-film related problems within the geometric measure theory community has progressed in several directions since the work outlined above. For some of these directions, see the other articles in this collection.

One additional major direction has been that of motion of surfaces, in particular motion by (weighted) mean curvature. For a survey of the physical applications and many mathematical approaches to weighted mean curvature in 1992 (there have been several more since then), see TAYLOR *et al.* (1992).

Note that motion by mean curvature is not the same as the motion that a soap bubble cluster would undergo as a result of air diffusing through the films from high pressure bubbles to lower pressure bubbles. For soap films, surface motion to acquire equal pressure within a bubble is fast compared with diffusion time, and hence solutions should have constant mean curvature on each interface at almost every time (with essentially instantaneous jumps when the topology of the froth changes). Motion by mean curvature, on the other hand, will almost never result in interfaces having constant mean curvature; it (or more generally, motion by weighted mean curvature) is appropriate for grain growth, where motion results from atoms detaching from one grain and reattaching to the grain on the other side of an interface. Weighted mean curvature should in fact be multiplied by a mobility factor M, which can also depend on the surface normal direction as well as possibly other factors, in order to get the normal velocity (M has units of distance over time, times volume over energy).

7.1. *Motion by (weighted) mean curvature*

The field of motion by mean curvature began with the publication of Brakke's book (BRAKKE, 1977). The book is full of many ideas and techniques; the subject began to be accessible by others only after the work of GAGE and HAMILTON (1986) on the simpler problem of the motion of a curve by its curvature. Much work has since been done in this PDE framework, on the motion both of curves and of hypersurfaces.

Within geometric measure theory and the calculus of variations, the problem was tackled again in ALMGREN *et al.* (1993), where a proof of all-time existence for boundaries of regions was given. Approximate flows for all time are defined by doing a sequence of minimizations for a given size time step, and a limit flow is found in the "flat" topology on integral currents as the time step goes to zero. This limit flow agrees with PDE flow wherever the PDE flow is known to have a solution (ALMGREN *et al.*, 1993) and was shown in ALMGREN and TAYLOR (1995) to agree with the previously defined motion by crystalline curvature for polygonal curves (ANGENENT and GURTIN, 1989; TAYLOR, 1991, 1993).

Motion of surfaces by crystalline weighted mean curvature has been partially studied (TAYLOR, 1992b).

The variational approach is being extended to general grain growth (including triple junctions) by CARABALLO (1996). But the computational aspect is more developed than the theoretical. Brakke's Surface Evolver (BRAKKE, 1992) computes grain growth in R^2 without operator intervention; in R^3 it still requires help. Motion of curves by crystalline curvature, including triple junctions and boundary points, was addressed in TAYLOR (1991, 1993).

7.2. *Other motion laws*

One way to evolve towards a soap froth configuration is to do motion by mean curvature with a velocity correction that ensures conservation of all enclosed volumes. This is essentially what Brakke's Surface Evolver does. Motion by weighted mean curvature is gradient flow for surface free energy in the L^2 inner product on normal velocity vector fields. There are other types of gradient flow for surface energy, especially if one imposes the constraint of enclosed fixed volume; see TAYLOR and CAHN (1994). For example, H^{-1} gradient flow for surface free energy results in surface diffusion. This is currently being investigated in geometric measure theory context: CHUNG (1996) has proved existence via a time-stepping approach similar to that described above (for the boundary of region). Crystalline surface diffusion in R^2 has also been defined and implemented computationally (CARTER *et al.*, 1995). One can combine both attachment kinetics and diffusion kinetics. The resulting motion is gradient flow in the $(1/D)H^{-1} + (1/M)L^2$ inner product, where M is the mobility and D is the surface diffusion constant. Sintering combines all of these types of motion, being primarily motion by weighted mean curvature where grains contact each other and primarily surface diffusion where grains contact vacuum or air. A general mathematical theory is lacking.

REFERENCES

ALLARD, W. K. (1972) On the first variation of a varifold, *Ann. Math.*, **95**, 417–491.

ALMGREN, F. (1965) *The Theory of Varifolds*, Mimeographed notes, Princeton University Mathematics Department.

ALMGREN, F. (1976) Existence and regularity almost everywhere of solutions to elliptic variational problems with constraints, *Memoir of the American Mathematical Society*, **165**.

ALMGREN, F. (1986) Deformations and multiple-valued functions, Geometric Measure Theory and the Calculus of Variations, *Proc. Symposia in Pure Math.*, **44**, 29–130.

ALMGREN, F. (1992) Questions and answers about area minimizing surfaces and geometric measure theory, *Differential Geometry, Proc. Symposia P. Math.*, American Mathematical Society, **54**, 29–53.

ALMGREN, F. and TAYLOR, J. E. (1976) The geometry of soap films and soap bubbles, *Scientific American*, **241**(1), 82–93.

ALMGREN, F. and TAYLOR, J. E. (1995) Flat flow is motion by crystalline curvature for curves with crystalline energies, *J. Differential Geometry*, **42**, 1–22.

ALMGREN, F., TAYLOR, J. E. and WANG, L. (1993) Curvature driven flows: A variational approach, *SIAM Journal of Control and Optimization*, **31**, 386–437.

ANGENENT, S. and GURTIN, M. E. (1989) Multiphase thermomechanics with interfacial structure. 2. Evolution of an isothermal interface, *Arch. Rational Mach. Anal.*, **108**, 323–391.

BOMBIERI, E. (1982) Regularity theory for almost minimal currents, *Arch. Rational Mech. Anal.*, **78**, 99–130.

BRAKKE, K. (1977) *The Motion of a Surface by its Mean Curvature*, Princeton University Press, Princeton.

BRAKKE, K. (1992) The surface evolver, *Experimental Mathematics*, **1**, 141–165.

BRAKKE, K. Soap films and covering spaces, *Journal of Geometric Analysis* (to appear).

CAHN, J. W. and TAYLOR, J. E. Thermodynamic driving forces and equilibrium in multicomponent systems with anisotropic surfaces (to appear).

CARABALLO, D. Ph.D. Thesis, Princeton University (in preparation).

CARTER, W. C., ROOSEN, A. R., CAHN, J. W. and TAYLOR, J. E. (1995) Shape evolution by surface diffusion and surface attachment limited kinetics on completely faceted surfaces, *Acta Metal. Mater.*, **43**(12), 4309–4323.

CHUNG, K. Ph.D. Thesis, Princeton University (in preparation).

DOUGLAS, J. (1931) Solution of the problem of Plateau, *Trans. A.M.S.*, **33**, 263–321.

GAGE, M. and HAMILTON, R. (1986) The heat equation shrinking convex plane curves, *J. Diff. Geom.*, **23**, 285–314.

GIBBS, J. W. (1928) *The Collected Works of J. W. Gibbs*, Vol. 1, Longmans, Green and Co., New York.

KINDERLEHRER, D., NIRENBERG, L. and SPRUCK, J. (1978) Regularity in elliptic free boundary problems I, *J. Analyse Math.*, **34**, 86–119.

LAWLOR, G. and MORGAN, F. (1994) Paired calibrations applied to soap films, immiscible fluids, and surfaces or networks minimizing other norms, *Pacific J. Math.*, **166**, 55–83.

MAXWELL, J. C., Capillary action, *Encyclopaedia Britannica*, 11th edition, **5**, 256–275.

MORGAN, F. (1995) *Geometric Measure Theory: A Beginner's Guide*, 2nd edition, Academic Press.

MORGAN, F. and TAYLOR, J. E. (1991) Destabilization of the tetrahedral point junction by positive triple junction line energy, *Scripta Metall. Mater.*, **25**, 1907–1910.

NITSCHE, J. C. C. (1977) The higher regularity of liquid edges in aggregates of minimal surfaces, *Nachr. Akad. Wiss. Göttingen Math.-Phys. Kl II*, 75–79.

OSSERMAN, R. (1970) A proof of the regularity almost everywhere of the classical solution to Plateau's problem, *Ann. Math.*, **91**, 550–569.

PLATEAU, J. A. F. (1873) *Statique Experimentale et Theorique des Liquides Soumis aux Seules Forces Moleculaires*, Gauthier Villiard, Paris.

RADO, T. (1933) *On the Problem of Plateau*, Julius Springer, Berlin.

TAYLOR, J. E. (1976) The structure of singularities in soap-bubble-like and soap-film-like minimal surfaces, *Ann. Math.*, **103**, 489–539.

TAYLOR, J. E. (1991) Motion by crystalline curvature, in *Computing Optimal Geometries* (ed. Jean E. Taylor), Selected Lectures in Mathematics, American Mathematical Society, 63–65 plus video.

TAYLOR, J. E. (1992a) Mean curvature and weighted mean curvature, *Acta Metall. Mater.*, **40**, 1475–1485.

TAYLOR, J. E. (1992b) Geometric crystal growth in 3D via faceted interfaces, in *Computational Crystal Growers Workshop* (ed. Jean E. Taylor), Selected Lectures in Mathematics, American Mathematical Society, 111–113 plus video 20:25–26:00.

TAYLOR, J. E. (1993) Motion of curves by crystalline curvature, including triple junctions and boundary points, *Differential Geometry, Proc. Symp. Pure Math.*, **51** (part 1), 417–438.

TAYLOR, J. E. (1994) book review of Wulff Construction, A Global Shape from Local Interaction, by R. Dobrushin, R. Kotecky and S. Shlosman, *Bull. Amer. Math. Soc.*, **31**, 291–296.

TAYLOR, J. E. and CAHN, J. W. (1994) Linking anisotropic sharp and diffuse surface motion laws via gradient flows, *J. Stat. Phys.*, **77**, 183–197.

TAYLOR, J. E., CAHN, J. W. and HANDWERKER, C. A. (1992) Geometric models of crystal growth, *Acta Metall. Mater.*, **40**, 1443–1474.

WHITE, B. (1985) Regularity of the singular sets in immiscible fluid interfaces and solutions to other Plateau-type problems, *Proc. Centre Math. Anal. Austral. Nat. Univ.*, **10**, 244–249.

A Counter-Example to Kelvin's Conjecture on Minimal Surfaces*

D. WEAIRE and R. PHELAN

Department of Pure and Applied Physics, Trinity College, Dublin 2, Ireland

Abstract. Kelvin's conjecture, that a b.c.c. arrangement of his minimal tetrakaidecahedron divides space into equal cells of minimum surface area, has stood for over one hundred years. We have found a counter-example, in the form of a structure analogous to that of some clathrate compounds and also related to the β-tungsten structure. Its surface area is approximately 0.3% less than that of Kelvin's structure.

In 1887 the *Philosophical Magazine* published Kelvin's classic analysis of the following problem (THOMSON, 1887). What space-filling arrangement of cells of equal volume has minimum surface area? This arises naturally in the theory of foams, when the liquid content is small. It has an obvious solution in two dimensions, but not in three.

Kelvin was able to draw on the rules of PLATEAU (1873) for equilibrium structures of this kind. The surfaces which bound the cells must meet at 120° and the lines which are formed by their intersections must meet at $\cos^{-1}(-1/3)$, the tetrahedral angle. He proposed the body-centred-cubic (b.c.c.) structure as a likely candidate for the optimal arrangement. What is now familiar as the Wigner-Seitz (or Voronoi) cell of the b.c.c. structure is an orthic tetrakaidecahedron constructed from six square faces and eight hexagons. Kelvin showed how a slight distortion of the hexagonal faces was sufficient to satisfy Plateau's rules. He remarked that "no shading could show satisfactorily the delicate curvature of the hexagonal faces". Figure 1 illustrates this structure. The curvature reduces the surface area of the cell by about 0.2% (PRINCEN and LEVINSON, 1987).

That this should be the optimal solution is an appealing conjecture, implicit rather than directly stated in Kelvin's original papers. It is rather like the proposition that the density of face-centred-cubic (f.c.c.) is the maximum achievable for equal spheres, but it rests on even less firm ground. The question has often been raised, as to whether Kelvin's choice can be bettered. For example, the discussions of WILLIAMS (1968), ROSS (1978) and PRINCEN and LEVINSON (1987) all make interesting contributions, but no-one has hitherto succeeded in proving or disproving the conjecture. Particularly entertaining is the review by the botanist MATZKE (1946) of various attempts to realize or observe the Kelvin

*Reproduced from *Philosophical Magazine Letters* (1994) Vol. **69**, No. 2, 107–110.

Keywords: Foam, Minimum Partitional Area, Space Division, Structure, Kelvin (given by the editors of FORMA).

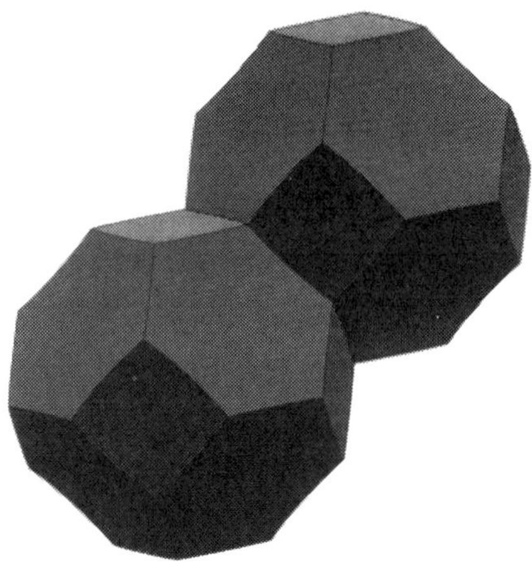

Fig. 1. Kelvin structure, constructed using the Surface Evolver package of BRAKKE (1992).

structure in real systems. As for the mathematicians, they have generally maintained an open mind, but no less an authority than WEYL (1952) gave it as his opinion that is was unlikely that Kelvin's construction could be improved upon.

Recently WEAIRE (1994) has offered some speculation on a somewhat wider issue, that of the optimal structures of foams of finite liquid content (wet foam). In continuation of that line of thought we have begun calculations of surface energies of rival structures for dry foam, particularly the clathrate structures suggested by WEAIRE (1994) as a natural choice for intermediate values of liquid content. Our first result has immediate and largely unexpected implications. It overthrows Kelvin's conjecture by providing a counter-example which has a significantly lower surface energy.

The structure in question is derived from the periodic cubic clathrate structure of Na_8Si_{46} described by KASPER *et al.* (1965). In this context, the cell vertices are tetrahedrally bonded Si atoms. These are entirely arranged in closed cages which enclose guest Na atoms, and can therefore be used to define a cellular structure. In adapting this simple clathrate to our purposes, we began with flat-sided polyhedral cells having the same topology. These were generated by a Voronoi construction using the Na positions as centres. The plane polyhedral cells generated by this initial construction do not have equal volumes. To equalize the volumes and produce the minimal structure we have used the "Surface Evolver" package of BRAKKE (1992). This program minimizes surface area, subject to the constraint of fixed cell volumes, for successively finer tesselations of the original cell faces. In this way the curvature of the minimal surface can be progressively approximated with increasing accuracy. In the present case, the topology of the cellular structure is undisturbed by this relaxation.

Figure 2 shows one unit cell of the resulting structure. The table compares its surface energy with that of the Kelvin structure and some others. For consistency with some

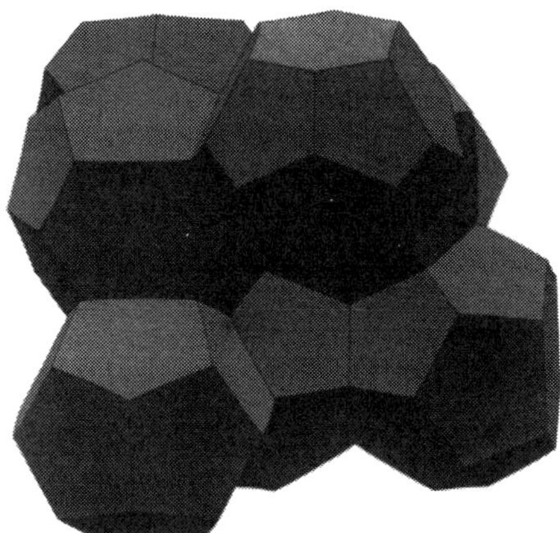

Fig. 2. Unit cell of the space-filling structure reported here, with surface area minimized. It consists of six 14-sided polyhedra and two 12-sided polyhedra, as described in the text.

Table 1. Isoperimetric quotient (figure of merit for area minimization) of various structures. An asterisk indicates non-space-filling cells.

Structure (lattice)	Isoperimetric quotient
Sphere*	1.000
Structure presented here (simple cubic lattice, eight cells)	0.764
Kelvin tetrakaidecahedron (b.c.c.)	0.757
Pentagonal dodecahedron*	0.7547
Orthic tetrakaidecahedron (b.c.c.)	0.7534
Rhombic dodecahedron (f.c.c.)	0.7405
Cube (s.c.)	0.5236

previous discussions we shall use as a figure of merit the so-called isoperimetric quotient defined by $36\pi V^2/A^3$. Here A is to be taken as the *average* cell surface area and V the cell volume. We see that the new structure has a surface energy which is approximately 0.3% less than that of Kelvin's solution, a remarkably large margin of superiority in this context.

Having established this result, we may rationalize it as follows. It has been repeatedly remarked that the ideal number of faces for a space-filling cell is close to fourteen: various geometrical/topological arguments point to this (WEAIRE and RIVIER, 1984). The 14-sided cell of least surface energy would appear to be that which consists of 12 pentagonal and two hexagonal faces. The clathrate structure which we have used consists mainly of such cells, with an admixture of one-third as many pentagonal dodecahedra. The 14-hedra are arranged in three mutually perpendicular, interlocking columns, the cells in each

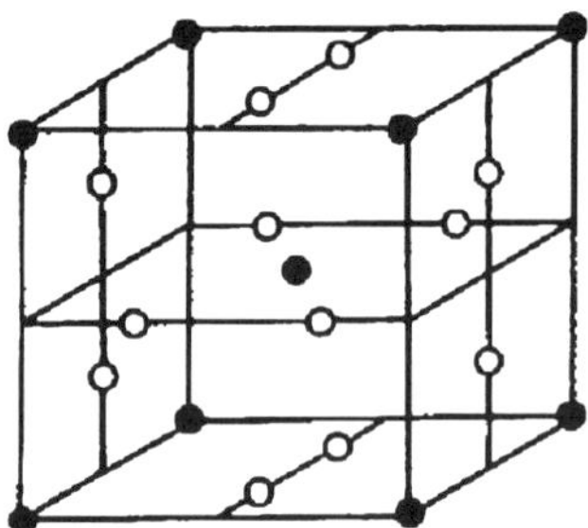

Fig. 3. The β-W structure, in which the atomic positions correspond to the cell centres of the structure depicted in Fig. 2.

column being joined by their hexagonal faces. The pentagonal dodecahedra lie between them on a b.c.c. lattice. The overall lattice periodicity is simple cubic.

Note that the cell faces of this structure are irregular. Only the hexagonal faces remain planar (since they lie in symmetry planes of the structure).

There is another, and for some purposes simpler, way of describing the structure. The cell centres are the atomic positions of the β-W structure illustrated in Fig. 3. This is one of the Frank-Kasper phases (see, for example, PEARSON, 1972). N. Rivier (private communication, 1993) has suggested that these should be of relevance to foam structures.

Of course, the structure discussed here can only be accorded the same provisional status as that previously enjoyed by Kelvin's, and it remains to be seen for how long it will hold the championship. In particular we have yet to check the other simple clathrate structure composed of pentagonal and hexagonal faces which arises in crystallography (WEAIRE, 1994), as it has a much larger unit cell. It will be possible in due course to make exhaustive computer searches, but the intuition which suggested the structure here presented tends to the view that it will be difficult to surpass.

To many the assertion that Kelvin's elegant and simple solution is not optimal will come as a surprise, and the subject, together with its experimental counterpart in the observation of foams and emulsions, is thereby rendered even more intriguing. Our own experimental observations will be presented in a further paper (WEAIRE and PHELAN, 1994).

Research supported by the EC SCIENCE Programme, contract SC*-CT92-0777 (DW) and Shell Research (Amsterdam): R.P. holds an Irish-American Partnership studentship. We would like to thank J. Sullivan and K. Brakke for advice on the Surface Evolver package and for checking the validity of this work. Thanks are also due to N. Rivier for much advice and N. Buttimore and T. Davis for their encouragement.

REFERENCES

BRAKKE, K. (1992) *Exp. Math.*, **1**, 141. The Evolver package, which is in the public domain, is available by anonymous FTP from *geom. umn. edu.*

KASPER, J., HAGENMULLER, P., POUCHARD, M. and CROS. C. (1965) *Science*, **150**, 1713.
MATZKE, E. (1946) *Am. J. Botany*, **32**, 130.
PEARSON, W. B. (1972) *The Crystal Chemistry and Physics of Metals and Alloys*, Wiley.
PLATEAU, J. (1873) *Statique Experimental et Theorique des Liquids Soumis aux Seules Forces Moleculaires*, Gauthier-Villars, Ghent, Paris.
PRINCEN, H. M. and LEVINSON, P. (1987) *J. Colloid Interface Sci.*, **120**, 172.
ROSS, S. (1978) *Am. J. Phys.*, **46**, 513.
THOMSON, W. (LORD KELVIN) (1887) *Phil. Mag.*, **24**, 503.
WEAIRE, D. (1994) *Phil. Mag. Lett.*, **69**, 99.
WEAIRE, D. and PHELAN, R. (1994) *Phil. Mag. Lett.*, **70**, 345–50.
WEAIRE, D. and RIVIER, N. (1984) *Contemp. Phys.*, **25**, 5.
WEYL, H. (1952) *Symmetry*, Princeton University Press, Princeton.
WILLIAMS, R. (1968) *Science*, **161**, 276.

The Structure of Monodisperse Foam*

D. WEAIRE and R. PHELAN

Department of Pure and Applied Physics, Trinity College, Dublin 2, Ireland

Abstract. We report various structural observations of monodisperse foams in the bulk, at surfaces, in cylinders and between glass plates. In general they tend to refute Matzke's assertion that even monodisperse foam is totally disordered. However, it appears that the tendency to order is quite limited in the bulk under normal circumstances. Small fragments are observed which correspond closely to the recently reported structure of Weaire and Phelan.

1. Introduction

We have recently reported calculations (WEAIRE and PHELAN, 1994b) which define an ideal structure for dry monodisperse foam, having a lower total (surface) energy than that of the Kelvin structure (THOMSON, 1887). These calculations were stimulated by certain experimental observations. In view of the interest in the new ideal structure for dry foam, it is desirable to place this work back in its experimental context.

In making and observing monodisperse foams (WEAIRE *et al.*, 1992, 1993) we have been repeatedly surprised by the rich variety of interesting effects which are readily observed. These have led us to believe that the classic work of MATZKE (1946) (see also MATZKE and NESTLER, 1946) is misleading. Its principal deficiency may be the effects of coarsening; in the key paper, measurements are described as having been made "usually on the same day or the following day". Our own experiments indicate that a scale of *tens of minutes* would have been more appropriate.

Our original goal was a better understanding of *bulk* structure (WEAIRE, 1994), but we have been repeatedly diverted by interesting observations of structure at *surfaces*, in thin *sandwiches* of bubbles, and in *narrow cylinders*. All these present highly ordered states without much special preparation while it appears the bulk does not.

---

*Reproduced from *Philosophical Magazine Letters* (1994) Vol. **70**, No. 5, 345–350.
Keywords: Foam, Structure, Surface, Honeycomb (given by the editors of FORMA).

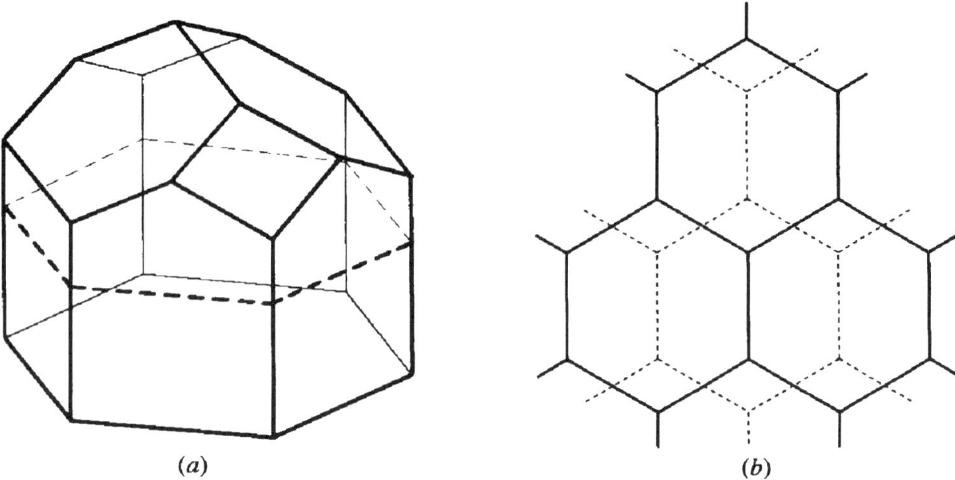

Fig. 1. (a) Sketch of a typical surface cell constructed simply from one half of a b.c.c. unit cell, cut on a (110) plane along the broken line, and with extended sides The surface lies below the figure. (b) Appearance of the surface structure generated by the cells shown in (a).

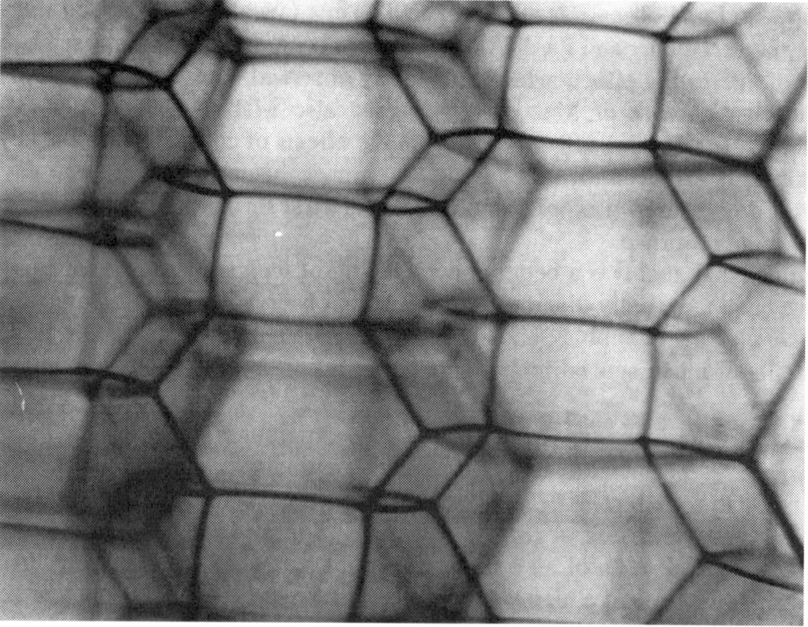

Fig. 2. A region of Kelvin cells in the second layer of bubbles of a large-diameter cylindrical sample. This structure is lost within the next three layers. (The bubble diameter in all the examples presented here is approximately 1.5 mm.)

2. Surfaces and Cylinders

Here we refer principally to the structure of monodisperse foam at a glass wall. Cell faces meet the wall at 90°. The end faces, lying in the surface itself, define a two-dimensional network of threefold vertices. Discounting occasional defects, this generally takes the form of an ordered hexagonal (honeycomb) structure. In contrast with this, Matzke found a highly disordered surface structure. It is true that he used a different procedure to make his monodisperse foams but it is so laborious as to defy tests of reproducibility.

In the case of cylindrical structures, the wrapping of a hexagonal network around a cylinder may be described and classified in a manner described by us (WEAIRE *et al.*, 1992) and now recognized as equivalent to the notation of the theory of phyllotaxis (ROTHEN and KOCH, 1989).

Narrow cylinders readily yield a variety of ordered internal structures consistent with the surface honeycomb pattern, including that mentioned by WEAIRE *et al.* (1992), with a core of Kelvin cells. We have found that this structure was previously reported by DODD (1955) who constructed it bubble by bubble in the manner of Matzke. We intend to categorize these structures, and the transitions between them, in further work.

In the case of a planar surface or a cylinder of large diameter, we generally observe an ordered periodic structure extending *at least* through one complete layer of cells. That is, not only do the fronts of these cells have an ordered periodic (honeycomb) structure, but also the backs of the cells present a uniform structure.

The surface cells are illustrated in Fig. 1. This corresponds precisely to the (110) surface structure of b.c.c. and therefore bears a close relationship to the Kelvin structure. It may be described as follows. Suppose that the ideal Kelvin structure is sliced in two on a (110) plane which bisects a layer of cells. These cells present a honeycomb structure, slightly strained with respect to a symmetric honeycomb. The volumes of these cells are easily readjusted to that of the original volumes by an extension of the cell walls, as also shown in Fig. 1. In this way the Kelvin structure is easily adapted to provide a low-energy surface structure, but only in the (110) direction. Indeed no other orientation is commonly observed.

The rival bulk structure which we have proposed (WEAIRE and PHELAN, 1994b) does not lend itself to such a construction. Hence it would appear that the Kelvin structure is favoured close to the surface. Indeed the second layer of bubbles often form Kelvin cells (Fig. 2). It is invariably found, however, that within three or four layers of the surface the Kelvin structure is lost. It should be plainly admitted that this is precisely the reverse of our initial prejudice in commencing this work.

3. Thin Sandwiches

By sandwich we mean a structure trapped between two glass plates. If only a single layer of bubbles is so enclosed, it exhibits the hexagonal honeycomb structure, as is now

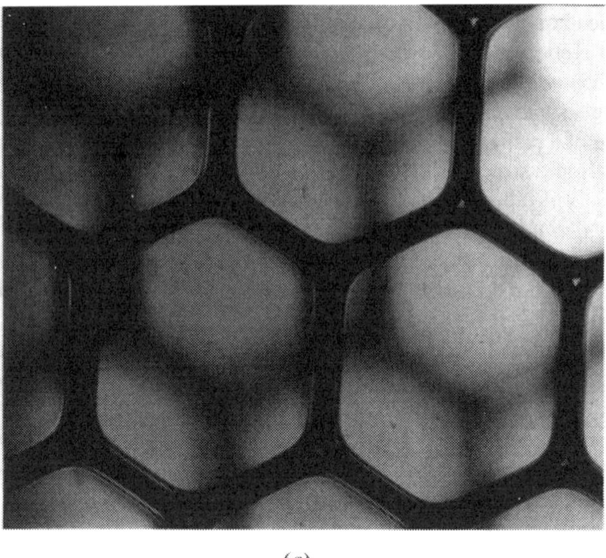

(a)

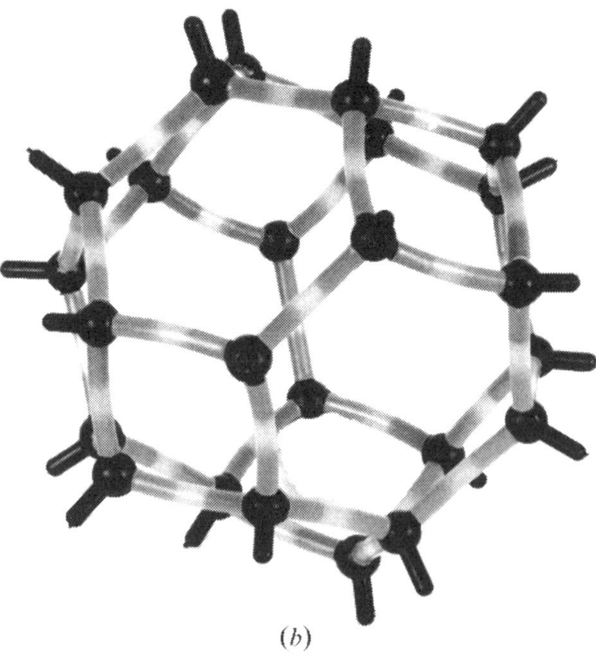

(b)

Fig. 3. (a) View of a double layer of dry foam (Toth structure). (b) Model of a "twisted" Kelvin cell. Note the relative orientation of the front and back portions. The intermediate region consists of two hexagons and four pentagons.

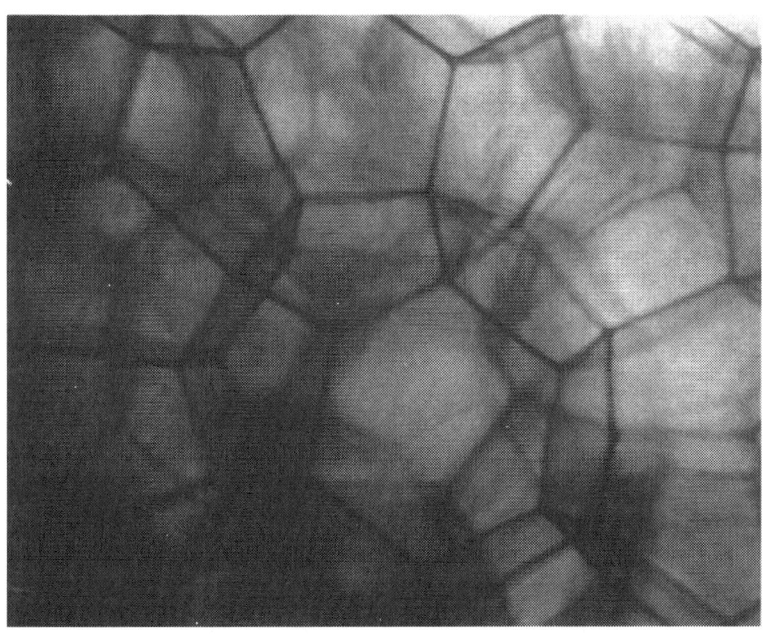

(a)

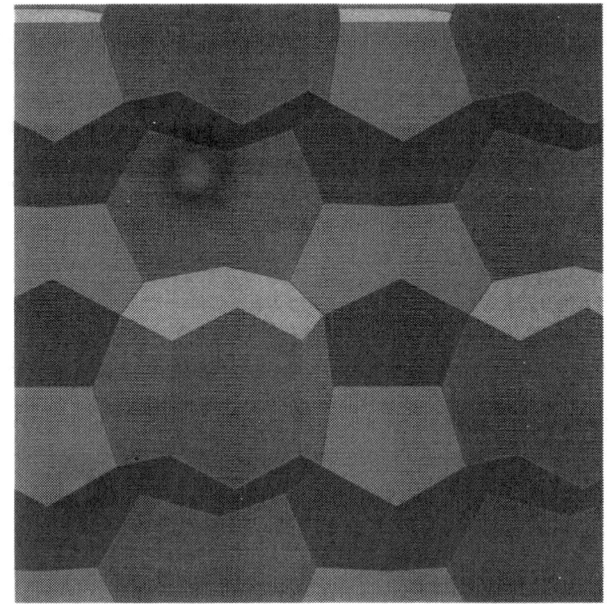

(b)

Fig. 4. (a) A region of bulk structure corresponding closely to the ideal foam structure described by WEAIRE and PHELAN (1994b). This picture was taken close to the centre of a cylindrical sample of 20 mm diameter. (b) A view of the ideal structure, generated using the Surface Evolver package (BRAKKE, 1992). Compare with (a).

familiar from the study of two-dimensional froths. If two layers are formed, then both are hexagonal. An interesting question already arises: what is the structure of the interface? Two b.c.c. (110) surfaces, back to back, provide the solution. This is indeed what is observed (Fig. 3(a)), as we have already noted elsewhere and further explored for wet foams (WEAIRE and PHELAN, 1994a). The result is in accord with the predictions of TOTH (1964), inspired by the study of the bee's honeycomb.

Thicker sandwiches exhibit more complex structures. For example, in the case of a three-layered sandwich a plane of Kelvin cells between two of the previously described surface layers is observed. However, a second, apparently metastable structure can also be constructed in which the two surface layers are not co-aligned. The central layer of bubbles can be described as "twisted" Kelvin cells, for which the front and back halves of each cell are rotated through 120° with respect to each other (Fig. 3(b)).

4. Bulk Structure

The central issue of the subject is the nature of the bulk structure. Are ordered structures to be found and, if so, which structures?

We have seen Kelvin cells, but these are close enough to the surface to be explicable as surface effects. We have also observed small fragments which appear to correspond to the new ideal structure reported by WEAIRE and PHELAN (1994b). Figure 4 shows an example of such a fragment with an accompanying matching view of the ideal structure.

We have not yet succeeded in finding regions which extend over more than a few cells. In future work we shall investigate techniques for "annealing" the structure in an attempt to enhance its order. Meanwhile it is interesting to note that the same topological structure has been observed in liquid crystals (CHARVOLIN and SADOC, 1988).

It would now appear that what was tentatively identified as a *bulk* structural transformation provoked by wetting in the work of WEAIRE *et al.* (1992) was predominantly the transformation of the surface cells. Ironically, this led to the re-examination of Kelvin's proposition for the bulk, and hence to the work which has finally refuted Kelvin's conjecture. The necessity for the present exposition, to clarify this tortuous course of events, should now be evident.

With the benefit of hindsight, it would appear that the subject has suffered in the past from an excess of theorizing and a shortage of experiments. This may be because Matzke's reports were misleading in stipulating such arduous experimental procedures, assembling structures bubble by bubble. The elementary (and old) procedure used by us, that of blowing bubbles under water, is easy and rapid. Now that computation of minimal surfaces is also straightforward, we can expect further rapid progress toward a full understanding of this system and its variety of interesting phenomena.

This research is supported by the EC Science Programme under Contract No. SC*-CT92-0777 (D.W.) and Shell Research (Amsterdam). R.P. holds an Irish-American Partnership studentship. We are grateful to J. Charvolin for drawing our attention to analogous liquid crystal structures, and to S. Findlay for assisting in some of the observations.

REFERENCES

BRAKKE, K. (1992) *Exp. Math.*, **1**, 141.
CHARVOLIN, J. and SADOC, J. F. (1988) *J. Phys., Paris*, **49**, 521.
DODD, J. D. (1955) *Am. J. Bot.*, **42**, 566.
MATZKE, E. B. (1946) *Am. J. Bot.*, **33**, 58.
MATZKE, E. B. (1950) *Bull. Torrey Bot. Club*, **77**, 222.
MATZKE, E. B. and NESTLER, J. (1946) *Am. J. Bot.*, **33**, 130.
ROTHEN, F. and KOCH, A. J. (1989) *J. Phys., Paris*, **50**, 633.
THOMSON, W. (Lord KELVIN) (1887) *Phil. Mag.*, **24**, 503.
TOTH, L. F. (1964) *Bull. Am. Math. Soc.*, **70**, 469.
WEAIRE, D. (1994) *Phil. Mag. Lett.*, **69**, 99.
WEAIRE, D., HUTZLER, S. and PITTET, N. (1992) *Forma*, **7**, 259.
WEAIRE, D. and PHELAN. R. (1994a) *Nature*, **367**, 123.
WEAIRE, D. and PHELAN. R. (1994b) *Phil. Mag. Lett.*, **69**, 107.
WEAIRE, D., PITTET, N., HUTZLER, S. and PARDAL, D. (1993) *Phys. Rev. Lett.*, **71**, 2670.

Organized Packing

N. RIVIER and T. ASTE

Laboratoire de Dynamique des Fluides Complexes, Université Louis Pasteur, 3 rue de l'Université, 67084 Strasbourg, France

Keywords: Packing, Foam, Entropy, Frustration, Disorder

1. Introduction

Filling three-dimensional Euclidean space regularly with cells is a compromise between best local packing and filling the whole space without gaps or overlap. The fact that large soap froths are almost always disordered demonstrates that this compromise is essential. There is simply no ordered structure whose energy is low enough to impose itself, without help from the boundary, even at low temperatures where the entropy (disorder) contribution to the free energy is negligible.

In this paper, we consider the problems of filling space by cells and of packing spherical atoms or grains. These two structures are dual of each other (atoms versus cells, edges versus faces, etc.). We analyse the problem of frustration between local packing and filling the whole space, and search for solutions by putting the whole, ordered structure in curved space (where it is not frustrated) and decurving progressively.

Experimentally, the packing solutions in Euclidean space are not ordered except for a remarkable collection of 24 known structures, called tetrahedrally closed-packed (t.c.p.) by metallurgists (Table 1). Their systematic enumeration has not yet been done. It is certainly not known whether the list is complete or not. In addition, all quasicrystals, which replace frustration by inflation symmetry (and end up with the non-crystallographic rotation symmetry of $2\pi/5$, 8, 10, or 12) are t.c.p.

Cells, bubbles in foams fill three dimensional, Euclidean space in an orderly fashion. Orderly does not necessarily mean ordered, but rather the result of some organizational principle, which we will ascribe to gentle decurving by local topological operations, of a structure which is initially curved.

A froth fills space with cells separated by interfaces. It is in equilibrium (apart from a slow coarsening due to diffusion). The cells are topological polyhedra and the interfaces, polygons. Stable mechanical equilibrium and topological stability under small local adjustments, require that exactly 3 faces meet on a common edge, 4 edges, 6 faces and 4 cells meet at a vertex. In soap froths, edges and interfaces are curved, interfaces meet at 120° on an edge, and edges at $\cos^{-1}(-1/3) = 109.47°$ at a vertex which constitutes a

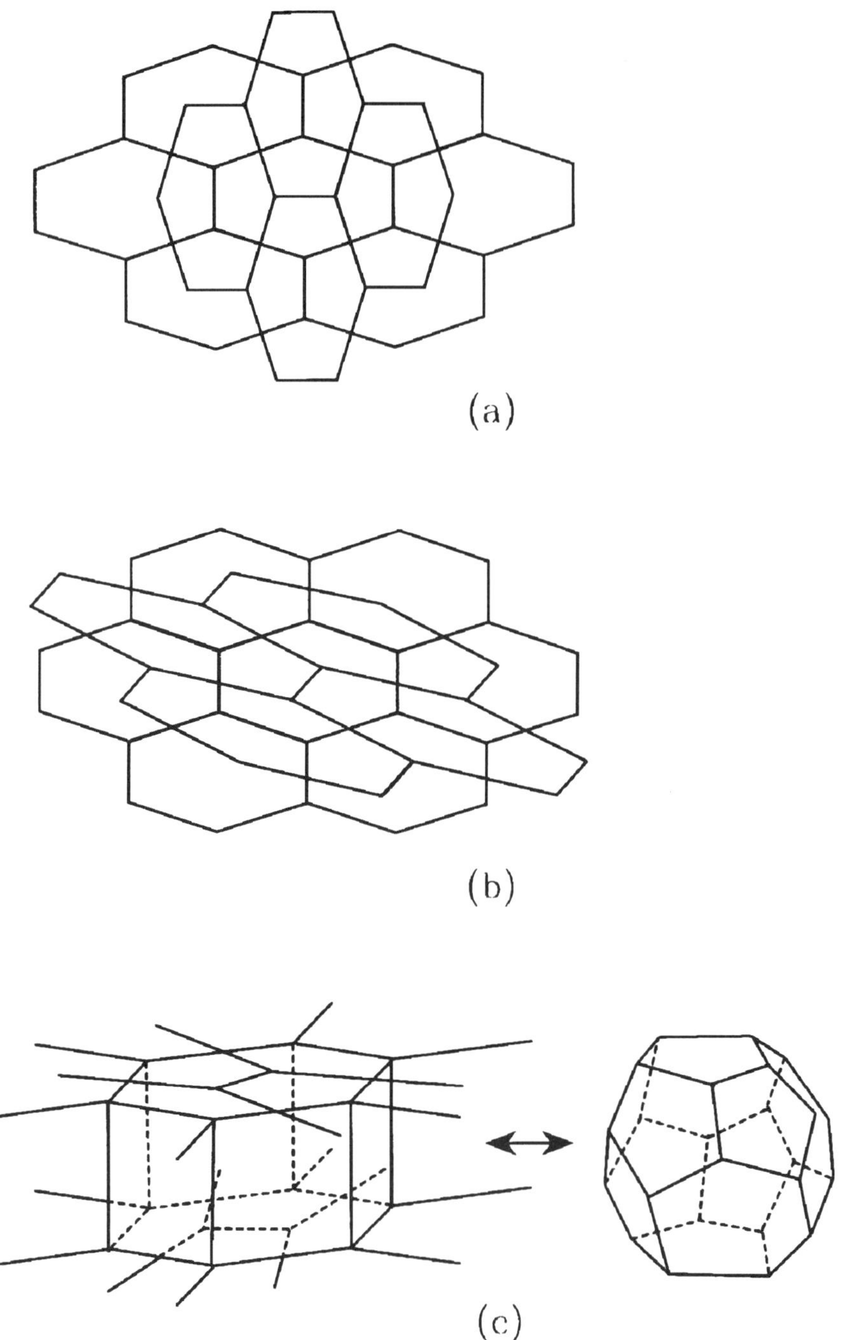

Fig. 1. (a), (b): Two consecutive Euclidean shells tiled with pentagonal tiles and the corresponding 3D Goldberg cell (c).

Table 1. The 24 known crystalline t.c.p. structures, adapted from SHOEMAKER and SHOEMAKER (1986). There are N atoms or Voronoi bubbles per unit cell, in the proportion p 16-sided, q 15-sided, r 14-sided and x dodecahedra. $\langle f \rangle$, the average number of facets of a bubble, also measures the energy of the disclination network (Eq. (1)).

		N	p	q	r	x	$\langle f \rangle$
A15	β-W, Cr_3Si	8	0	0	3	1	13.5
σ	β-U, $Cr_{46}Fe_{54}$	30	0	2	8	5	13.47
H	complex	30	0	2	8	5	13.47
K	complex	82	0	6	21	14	13.46
F	complex	52	0	4	13	9	13.46
J	complex	22	0	2	5	4	13.45
ν	$Mn_{81.5}Si_{18.5}$	186	6	10	40	37	13.44
Z	Zr_4Al_3	7	0	2	2	3	13.43
P	$Mo_{42}Cr_{18}Ni_{40}$	56	1	2	5	6	13.43
δ	MoNi	56	1	2	5	6	13.43
K	$Mn_{77}Fe_4Si_{19}$	220	7	4	19	25	13.42
R	$Mo_{31}Cr_{51}Co_{18}$	159	8	6	12	27	13.40
μ	Mo_6Co_7	39	2	2	2	7	13.38
—	K_7Cs_6	26	2	2	2	7	13.38
pσ	$V_6(Fe,Si)_7$	26	2	2	2	7	13.38
M	$Nb_{48}Ni_{39}Al_{13}$	52	2	2	2	7	13.38
I	$V_{41}Ni_{36}Si_{23}$	228	4	2	2	11	13.37
C	$V_2(Co,Si)_3$	50	6	2	2	15	13.36
T	$Mg_{32}(Zn,Al)_{49}$	162	20	6	6	49	13.36
SM	$Mg_{32}(Zn,Al)_{49}$	162	46	0	18	98	13.36
X	$Mg_{45}Co_{40}Si_{15}$	74	10	2	2	23	13.35
—	Mg_4Zn_7	110	16	2	2	35	13.35
C14	$MgZn_2$	12	1	0	0	2	13.33
C15	$MgCu_2$	24	1	0	0	2	13.33

perfect tetrapod. The equilibrium of the soap froth is only mechanical. Slow diffusion of air between bubbles leads to a coarsening of the froth and to a reduction of the total number of the bubbles. Nevertheless, the soap froth structure corresponds to a minimum in the (large) elastic energy contained in the interfaces.

Froths can also be regarded as the dual structures of packing of atoms (or convex grains). The dual requirement to 4 cells meeting at a vertex is that 4 atoms should pack together in a tetrahedron, which for a large class of metallic alloys should be as regular as possible. These atomic structures are known as tetrahedrally closed-packed phases (t.c.p.), or Frank-Kasper phases (although not all t.c.p are FK). There are 24 known ordered metallic alloy t.c.p. (SHOEMAKER and SHOEMAKER, 1986) (see Table 1), in addition to all quasicrystals.

Froths are usually disordered because they are geometrically frustrated and also because the large entropy easily offsets the small energy cost.

The existence of an ordered soap froth had been proposed a hundred years ago by Lord Kelvin (THOMSON, 1887). Kelvin's bubbles are all 14-sided truncated octahedra and they pack in a body-centred cubic lattice, which is not a t.c.p. Recently, Weaire and Phelan (WP) (WEAIRE and PHELAN, 1994) have found that a simple-cubic arrangement of 6 Goldberg cells (GOLDBERG, 1934) (14-sided polyhedra, with 2 hexagons at the base and the top, and 12 pentagons on the side, see Fig. 1) and 2 pentagonal dodecahedras had 0.3% less interfacial area per unit volume than Kelvin's. The WP structure is in fact a t.c.p., called A15 or β-tungsten structure. In the A15, the volumes of the two bubbles are slightly different.

Ordered arrangements of bubbles have also been observed in a limited environment (cylindrical tube) (WEAIRE *et al.*, 1992, 1993; PITTET *et al.*, 1995, 1996). Bulk foams have resolutely remained disordered. Nevertheless BIBETTE *et al.* (1990, 1992) have succeeded in making an ordered emulsion after several creaming processes (to concentrate and make the emulsion more monodisperse).

Natural opals are compact, ordered packings of silicate spheres. Ordinary opals arrange in fcc or hcp, which are not t.c.p. However, natural opals with the Laves structure (the t.c.p. C15) have been found recently (SANDERS, 1980; CASEIRO *et al.*, 1994), with silicate spheres of two different sizes.

2. From Curved Space to Real Cellular Structures

The structures of foams or of their dual packings can be built algorithmically, by random sequences of decurving operations (SADOC and RIVIER, 1987).

The closest packed local arrangement of four spheres is a regular tetrahedron. If one imposes that the tetrahedron is as regular as possible (corresponding to closest packing of 4 atoms), its dihedral angle is approximately $2\pi/5.1$. Thus, 5.1 regular tetrahedra fit about one edge (like the segments of an orange). This is doubly frustrating: The decimal fraction wastes space between atoms, and 2π over the nearest 5 is not a crystallographic rotation. (The corresponding frustration in the froth is expressed by the fact that isotropic bubbles of equal sizes have 13.4 interfaces, far from the ordered froth of Kelvin.)

Frustration is relieved simply by embedding the structure in positively curved space (KLÉMAN and SADOC, 1979). Positive curvature takes care of the 0.1 (the lips, slightly ajar in Euclidean space, are just closed by curvature), and the five-fold symmetry is crystallographic in the finite curved space. The resulting structure is a scaffolding for the hypersphere S^3, polytope {3,3,5} (COXETER, 1973), consisting of 120 spheres (vertices), 600 regular tetrahedra, 5 of which are incident on each of the 720 edges, and 20 on each vertex, making up an icosahedral coordination shell. Polytope {3,3,5} is therefore the "ideal" close-packed structure. Its dual, polytope {5,3,3}, is a regular froth of 120 dodecahedral cells, with all their interfaces pentagonal. But it is a curved space ideal, which materializes as various structures in Euclidean space (some ordered, some random). To obtain these Euclidean structures, one must either project (RIVIER and LAWRENCE, 1985), or, better, decurve iteratively, without (SADOC and MOSSERI, 1985) or with disorder (SADOC and RIVIER, 1987).

The Euclidean structure is given by sequences of decurving operations (SADOC and RIVIER, 1987). Decurving leaves a skeleton of negative disclination lines behind. Negative disclinations segments are edges of the packing with 6 (instead of 5) incident tetrahedra. In the dual froth, they are 6 (instead of 5)-sided faces. The disclination segments form uninterrupted lines.

Iterative decurving will preserve the local icosahedral environment as much as possible. It is done in sequences which can be either random but spatially uniform, or are also random in space (SADOC and RIVIER, 1987). A product of decurving operations is a decurving operation.

Most of the physical properties (energy, density) of the materials are independent of the order of decurving operations of the sequence. Defects, inhomogeneities are commutators between decurving operations which are more localized if they occur later in the sequence. Conversely, an early commutator gives rise to an extended defect.

3. Energy of the Cellular Structures

Polytope {3,3,5} is the closest packing, in curved space, of equi-sized spheres. Its dual {5,3,3} (which is topologically a froth), fills best curved space with (dodecahedral) bubbles of equal volumes. Polytope {5,3,3} is the froth of lowest energy. It is ordered, and is defined in a positively curved space, which is finite.

Consequently, a real, infinite structure in Euclidean space will have two, and possibly three contributions to its energy.

i) A topological contribution to *decurve* S^3 into Euclidean space, carried by *negative disclination lines*.

ii) The elastic energy necessary to impose equal volumes.

iii) In addition the decurving process may generate local *defects*, which are mechanically and topologically stable "wool-balls" of positive and negative disclinations (SADOC and RIVIER, 1987). They are local, stable inclusions, signatures of spatial disorder (RIVIER and SADOC, 1988a). But they also occur in some ordered structures like fcc and the Kelvin froth.

The first two contributions are essential and inescapable in three dimension.

Clearly, the structures of lowest energy have no defects and their total energy is a combination of decurving (i) and elastic (ii) contributions. The experimentally known structures without defects and positive disclinations are the t.c.p. They all have an average number of sides $\langle f \rangle$ between 13.333... and 13.5 (Table 1), which surrounds the ideal value of 13.40 (RIVIER, 1982). Their structure is entirely characterized by a network of negative disclinations (the major skeleton of Frank and Kasper (FRANK and KASPER, 1959; SHOEMAKER and SHOEMAKER, 1986).

It has not been possible to obtain, theoretically or experimentally, Euclidean structures outside the t.c.p. range of $\langle f \rangle$ which are free of positive disclinations. The t.c.p. range is bounded by the Laves phase (C15) ($\langle f \rangle = 13.333...$) and the A15 ($\langle f \rangle = 13.5$).

The C15 can easily be obtained by systematic decurving operations (SADOC and MOSSERI, 1985). It has the shortest disclination network of all t.c.p. (with 4 disclinations incident on every vertex at 109.47°, the solution of the Steiner problem in 3D (HILDEBRANDT and TROMBA, 1984). To get a lower value for $\langle f \rangle$, one would have to include cells of very different sizes and put in positive disclinations (4- and 3-sided faces).

By contrast, it has not yet been possible to obtain the A15 by iterative decurving of polytope {3,3,5}, although its construction in one decurving step is elementary (RIVIER and SADOC, 1988b). Here the disclinations are straight lines across the material. Again, the total length is optimized. One can attempt to arrange the disclination lines differently as, for example, in the structure made of Euclidean sheets of Goldberg cells only, constructed in ASTE *et al.* (1996). However, this structure lies in negatively curved space, as its characteristic hyperbolic distortion (elongation in one direction, shrinkage in an other) indicates, see Fig. 1. We have gone overboard in decurving. The upper bound of $\langle f \rangle = 13.5$ for structures without positive disclinations is therefore only a strong conjecture. Incidentally, the two extreme structures bounding the t.c.p. range (C15, A15, see Table 1) are the simplest t.c.p. We will see that they also have the lowest energy among the t.c.p.

We now construct the total energy.

(i) The decurving energy is proportional to the topological length of the disclination network. Because this network consists only of 2-, 3- and 4-valent nodes, as befits a stable network of lines under tension (higher nodes can split into two or more canonical nodes and decrease the total length of the network), the decurving energy per line depends linearly on $\langle f \rangle$

$$\frac{E_{discl}}{N} = \frac{1}{2} t \frac{(2n_{14} + 3n_{15} + 4n_{16})}{N} = \frac{1}{2} t (\langle f \rangle - 12), \tag{1}$$

since every disclination edge joins two cells, and 14-, 15-, and 16-sided cells have two, three and four incident disclination segments, respectively. Here n_{14}, n_{15} and n_{16} are the number of 14-, 15- and 16-sided bubbles and $N = n_{12} + n_{14} + n_{15} + n_{16}$ is the number of cells per crystallographic unit cell. The parameter t is the work against tension to stretch a disclination line.

(ii) The elastic energy is less obviously a function of $\langle f \rangle$ only, but this can be justified if all the cells are nearly equisized and isotropic (RIVIER, 1982). It is then a quadratic function in $\langle f \rangle$, with an extremum at some value b. This extremum is in fact a *maximum*, and not a minimum as suggested in RIVIER (1994), for the following reason: Any fluctuation about the extremum leads to diffusion of air from the smaller to the larger bubble and to a slow coarsening of the structure. A froth made with equisized bubbles is therefore in *unstable* equilibrium. Thus, the elastic energy per cell is

$$\frac{E_{vol}}{N} = -\frac{1}{2} \eta (\langle f \rangle - b)^2. \tag{2}$$

The total energy is therefore quadratic in $\langle f \rangle$,

$$\frac{E_{tot}}{N} = -\frac{1}{2}\eta\left(\left(\langle f\rangle - f_0\right)^2 - \frac{t}{\eta}\left(-\frac{1}{4}\frac{t}{\eta} + f_0 - 12\right)\right) \tag{3}$$

and has a maximum at the compromise value

$$f_0 = b + \frac{1}{2}\frac{t}{\eta} = 13.40. \tag{4}$$

Given that any structure with $\langle f\rangle$ outside the t.c.p. range will have a large additional contribution of the defects to its energy, the structures of lowest energy will be the A15 (WP) and the Laves phase (C15). These two simplest t.c.p. had been suggested by CHARVOLIN and SADOC (1988) as possible micellar structures in amphiphilic liquid crystals. Although the two structures have lowest total energy, neither are energy minima. They are just at the two extremities of the range of t.c.p. free of positive disclinations. Moreover, the A15 energy is lower than the Laves phase. Phelan (private communication)

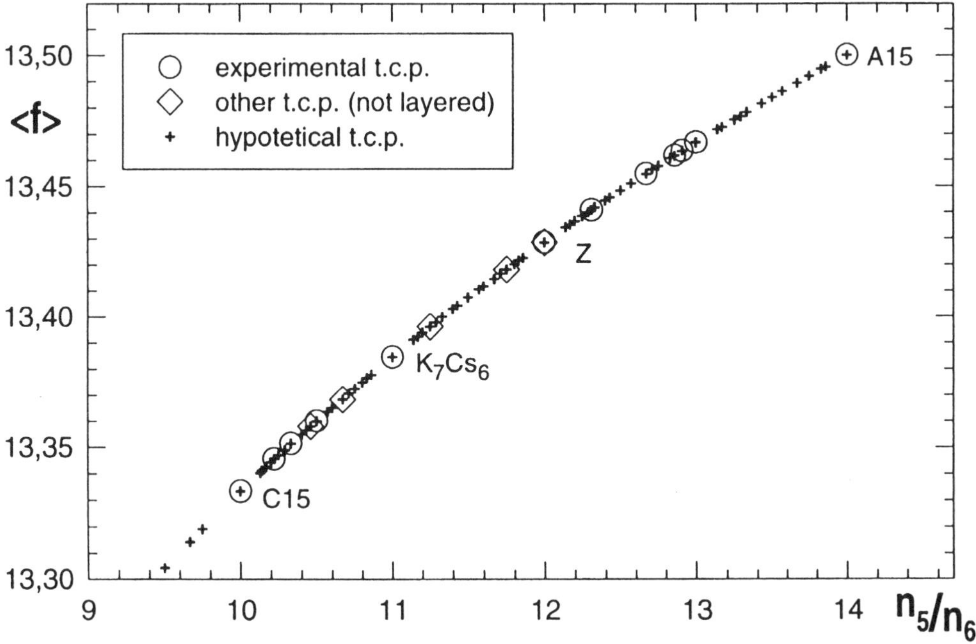

Fig. 2. The average number of neighbours ($\langle f\rangle$) as function of the proportion of hexagons (n_6) and pentagons (n_5) in the shell of t.c.p. structures. The tendance to gather into distinct groups may indicate either the existence of unfavourable configurations or structural mode-locking into the simplest t.c.p. (A15, K_7Cs6, Z, C15).

has obtained a relative improvement of about 0.14% (different bubble sizes, 0.8% if equi-sized bubbles) for the A15 over the Laves phase, which leads to the reasonable ratio $t/\eta \simeq 7.7$ (of the two essential, conflicting contributions to the energy decurving plays the leading role). All 24 t.c.p. have relative energies within 0.25% above the A15. This value is very small, smaller even than the improvement of 0.3% of WP over the Kelvin structure. Indeed, ordered foams occur extremely rarely.

In Fig. 2 are plotted the values $\langle f \rangle$ of the average number of neighbours per cell for all the 24 t.c.p., as well as for hypothetical t.c.p. constructed by succession of shells, with simplest integer numbers n_5 and n_6 of 5 and 6-sided faces in each shell (ASTE *et al.*, 1996).

Incidentally we can also find a hypothetical t.c.p. which maximizes the orientational entropy. It would have $\langle f \rangle = 13.29$ ... (ASTE *et al.*, 1996). The structure of natural froths is a compromize between the minimization of the configurational energy and the maximization of the entropy. Therefore we can predict that, in natural structures, $\langle f \rangle$ lies between 13.29... and 13.5. The lower bound corresponds to configurations with maximal orientational entropy, whereas the upper bound corresponds to minimal interfacial energy. All the known t.c.p. structures falls within this range.

4. Conclusions

Froths and packings are compromizes between closest local packing and space-filling. The best compromizes are obtained from decurving the spherical polytope {3,3,5} or its dual froth {5,3,3}. This produces a range of structures which include the t.c.p. We have shown here that the structures with the lowest energies are the two t.c.p. at the extremities of this range, namely A15 and C15 (Table 1). The A15 has the lowest energy and it is also the ideal froth of WEAIRE and PHELAN (1994). All the other t.c.p. have relative energies higher by less than 0.25%.

We acknowledge many discussions with D. Boosé, N. Pittet, D. Weaire, R. Phelan, J. F. Sadoc and J. Charvolin. This work was partially supported by EU, HCM Program, "FOAMPHYS" network, contract EBRCHRXCT940542.

REFERENCES

ASTE, T., BOOSE, D. and RIVIER, N. (1996) *Phys. Rev. E*, **53**, 6181–6191.
BIBETTE, J., ROUX, D. and NALLET, F. (1990) *Phys. Rev. Lett.*, **65**, 2470–2473.
BIBETTE, J., MORSE, D. C., WITTEN, T. A. and WEITZ, D. A. (1992) *Phys. Rev. Lett.*, **69**, 2439.
CASEIRO, J., RANTSORDAS, S. and GAUTHIER, J.-P. (1994) *Proceedings of SFP-JMC4*, Rennes, 772.
CHARVOLIN, J. and SADOC, J. F. (1988) *J. Physique*, **49**, 521.
COXETER, H. M. S. (1973) *Regular Polytopes*, Dover, New York.
FRANK, F. C. and KASPER, J. S. (1959) *Acta Crystallogr.*, **12**, 483.
GOLDBERG, M. (1934) *Tohoku Math. J.*, **40**, 226.
HILDEBRANDT, S. and TROMBA, A. (1984) *Mathematics and Optimal Form*, Scientific American.
KLÉMAN, M. and SADOC, J. F. (1979) *J. Physique Lett.*, **40**, 569.
PITTET, N., RIVIER, N. and WEAIRE, D. (1995) *FORMA*, **10**, 65–73.
PITTET, N., BOLTENHAGEN, P., RIVIER, N. and WEAIRE, D. (1996) *Europhys. Letters*, **35**, 547–552.

RIVIER, N. (1982) *J. Physique Coll.*, **43**, C9-91.
RIVIER, N. (1994) *Phil. Mag. Lett.*, **69**, 297–303.
RIVIER, N. and LAWRENCE, A. J. A. (1985) *J. Physique (Coll).*, **46**, C8-409.
RIVIER, N. and SADOC, J. F. (1988a) *J. non Cryst. Solids*, **106**, 278–281.
RIVIER, N. and SADOC, J. F. (1988b) *Europhys. Lett.*, **7**, 523.
SADOC, J. F. and MOSSERI, R. (1985) *J. Physique*, **46**, 1809.
SADOC, J. F. and RIVIER, N. (1987) *Phil. Mag. B*, **55**, 537.
SANDERS, J. V. (1980) *Phil. Mag. A*, **42**, 705–720.
SHOEMAKER, D. P. and SHOEMAKER, C. B. (1986) *Acta Crystallogr.*, **B42**, 3.
THOMSON, W. (Lord Kelvin) (1887) *Proc. Roy. Soc.*, **55**, 1.
WEAIRE, D. and PHELAN, R. (1994) *Phil. Mag. Lett.*, **69**, 107.
WEAIRE, D., HUTZLER, S. and PITTET, N. (1992) *FORMA*, **7**, 259–263.
WEAIRE, D., PITTET, N., HUTZLER, S. and PARDAL, R. (1993) *Phys. Rev. Lett.*, **71**, 2670.

Comparing the Weaire-Phelan Equal-Volume Foam to Kelvin's Foam

R. KUSNER[1] and J. M. SULLIVAN[2]

[1]*Department of Mathematics, University of Massachusetts, Amherst, MA 01003, U.S.A.*
[2]*School of Mathematics, University of Minnesota, Minneapolis, MN 55455, U.S.A.*

Keywords: Foams, Soap Films, Minimal Surfaces, Partitions, TCP Structures

1. Introduction

The problem of partitioning space into equal-volume cells, using the least interface area, was considered in 1887 by Sir William THOMSON, Lord KELVIN (1887). His proposed solution yields a foam with cells of a single shape, tiling space by the translations of the body-centered cubic lattice. In 1993, Denis WEAIRE and Robert PHELAN (1994) proposed a new equal-volume foam with two different cell shapes, which uses less area according to their computer experiments with Ken BRAKKE's Evolver (1992).

Mathematically, one difficulty is that neither foam can be described explicitly; even their existence is troublesome. In this note, we will examine these two foams, showing that the Weaire-Phelan foam is in fact more efficient than the Kelvin foam, and we will describe some other interesting candidate foams. More mathematical details, including a proof of existence for the Kelvin foam, are forthcoming in ALMGREN *et al.* (in preparation).

A *partition* of space is a division of $\mathbb{R}^3$ into disjoint cells. We are mainly interested in the surfaces forming the interface between the cells. The partitions we consider will be periodic with respect to some lattice, with some number n of cells in each periodic domain; thus they could be viewed as partitions of a quotient three-torus into n cells. Fixing the volume of each cell to be V, we want to minimize the *cost* of the equal-volume partition, defined scale-invariantly to be $\mu := A^3/V^2$, where A is the average interface area per cell. (Note that A is really half the boundary area of a typical cell, since each interface is shared by two cells.) If we scale to make $V = 1$, the case of unit-volume cells, then our cost is simply the cube of the total surface area in a periodic domain, divided by n^3.

If we start with some partition, perhaps polyhedral, and let it relax until the cost is a minimum, at least among nearby partitions, then the resulting stable partition should exhibit the geometry of a cluster of soap bubbles. Thus a stable partition should follow the rules recorded by PLATEAU in 1873 for such clusters, and proved by Jean TAYLOR in 1976 for a certain mathematical model (due to Fred Almgren) of compound bubbles.

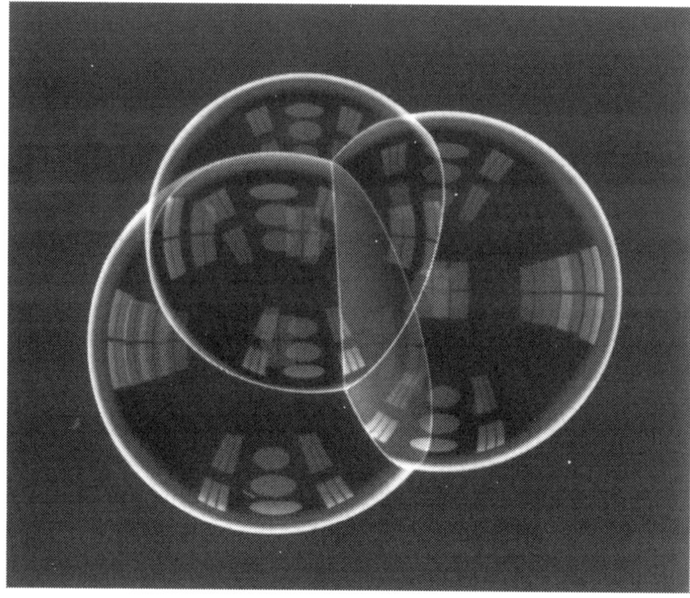

Fig. 1. This triple bubble in space exhibits the Plateau singularities: smooth surfaces meet along triple-junction curves at 120° angles, and these junctions come together at tetrahedral points.

These *Plateau rules* state that the interfaces are smooth surfaces of constant mean curvature, except where they meet in threes (at equal 120° angles) along smooth arcs. These arcs are, furthermore, allowed to come together (four at a time) at isolated points, where the configuration is tetrahedral; but no other singularities are allowed. The allowed singularities are observed for instance in a cluster of three or more soap bubbles, as in Fig. 1. The mean curvature of each surface is simply the pressure difference between the two cells on either side; this implies for instance that the three mean curvatures around a triple junction sum to zero.

We can view these rules as having a geometric part (about curvatures and equal angles) and a combinatorial part: that cells should meet along edges in threes, and at corners in fours, tetrahedrally. This combinatorics is the same as is observed generically in Voronoi partitions. Given a collection of sites in space, we define the *Voronoi cell* (SENECHAL, 1990; OKABE *et al.*, 1992) for each site to be the region consisting of points closer to that site than to any other. Figure 2 shows an example in the plane. In space, the boundary between two adjacent cells in the partition is a polygonal piece of the plane perpendicularly bisecting the segment between the two sites. These bounding polygons meet along segments equidistant from three sites, which terminate at vertices equidistant from four sites. The Voronoi partition of a symmetric collection of sites will share its symmetry.

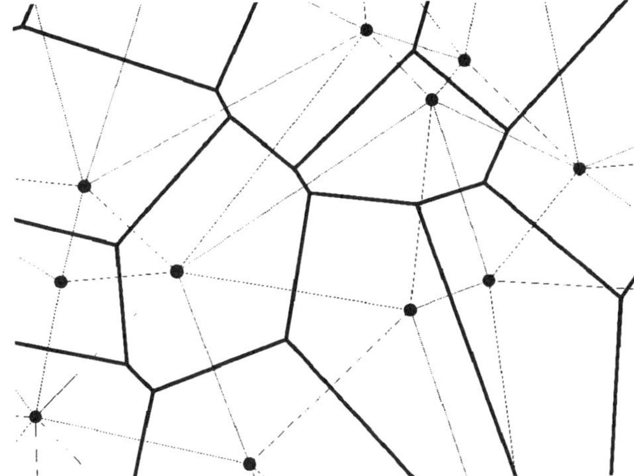

Fig. 2. Sites in the plane (marked with dots) together with their Voronoi partition (heavy lines). The thin lines are the dual Delone triangulation, connecting sites with adjacent Voronoi cells.

This similarity in combinatorial structure suggests we might find foams as relaxed Voronoi partitions. We specify certain sites and compute their Voronoi partition to get polyhedral cells in the right combinatorial structure; then we can follow mean curvature flow (with volume constraints) toward a stationary foam satisfying the geometric Plateau rules. This idea was implemented numerically at the Geometry Supercomputer Project in 1988, using Sullivan's VCS software (SULLIVAN, 1988) to compute Voronoi partitions in three dimensions, and Brakke's Evolver to relax them to foams. Some results appeared in the movie "Computing Soap Films and Crystals", produced by Almgren, Brakke, Sullivan and Taylor, and published in ALMGREN (1991). Those experiments found no foams better than Kelvin's. But Weaire and Phelan knew of an interesting pattern (often called A15) arising in chemical clathrates, and used the same pair of programs to find that the corresponding foam did have lower cost.

There is not enough mathematical theory for mean curvature flow with triple junctions to know in general when this Voronoi procedure for constructing foams really works. Certainly topological or combinatorial changes may occur if the relaxation flow proceeds far enough; such singularities can lead to nonuniqueness for the continued mathematical flow (ANGENENT *et al.*, 1995). However, all the foams we will consider will be constructed in this way, evidently without such topological changes. In each case, the Voronoi sites determine a certain pattern of combinatorics and symmetry, and presumably there is a unique relaxed foam in this pattern, close to the one we observe computationally.

The cells in a Voronoi partition do not in general have equal volumes. The Evolver can adjust the volumes towards the desired targets as the numerical relaxation proceeds. But mathematically it is helpful to start with the correct volumes. A *weighted* Voronoi partition is a generalization in which different weights on the various sites change the relative sizes of the cells; each interface plane is moved away from the site of higher weight. With appropriate weights, the initial Voronoi cells will have the desired volumes, and there is less chance that the mean curvature flow will cause combinatorial changes.

2. The Kelvin Foam

Kelvin described his foam as a relaxation of the Voronoi partition for the body-centered cubic lattice BCC, whose cells are congruent truncated octahedra. We will take this BCC lattice to be generated by cyc(2,2,–2), by which we mean the three vectors given by cyclic permutations of these coordinates. This BCC lattice consists of the cubic lattice CUB $= 4\mathbb{Z}^3$ and its translate by (2,2,2). The Voronoi cell around the origin touches fourteen others: the six nearest neighbors in the cubic lattice, and eight along body diagonals. (Kelvin called this cell a tetrakaidecahedron because of its fourteen faces. Of course, there are many other 14-hedra, such as the one we encounter in the next section; this cell is better described as the (Archimedean) truncated octahedron.) The 24 vertices of this cell are the points cyc(0, ±1, ±2) and cyc(±2, ±1, 0). This unrelaxed foam has cost

$$\mu = \frac{27}{128}\left(30\sqrt{3} + 37\right) = 18.7653\cdots.$$

The Voronoi partition of course has the full group of symmetries of the lattice. We will call this group G_0; it is the crystallographic group $Im3$. It is generated by the BCC translations and by sign changes and permutations of the coordinates. This group acts transitively on (oriented) edges of the Voronoi partition (or the Kelvin foam). In particular, all the edges are congruent.

We will say, of any partition which has this same combinatorics and symmetry, that it is in the *Kelvin pattern*. In particular, when we relax the polyhedral partition, the symmetry is preserved, so the resulting Kelvin foam is in the same pattern. As the foam relaxes, the vertices are fixed by symmetry. The square faces remain in their mirror planes, although the edges bow out within these planes. The diagonals of the hexagons remain fixed along axes of rotational symmetry, while the hexagons become shaped like monkey-saddles. (The diagonals of the squares also lie along axes of two-fold rotational symmetry.)

Physically, it is easy to see that soap film in the Kelvin pattern will relax (moving only slightly, as just described) to the foam shown in Fig. 3(a). Mathematically, we can also prove the existence and uniqueness of such a foam (ALMGREN *et al.*, in preparation).

Consider now the piece of the foam nearest to a particular triple-junction edge (as in Fig. 3(b)). It includes a quarter of one square and sixths of two hexagons, and its boundary is along axes of 180° rotational symmetry. This unit will generate the entire foam under these rotations; it also has further four-fold symmetry given by a pair of mirror planes. The boundary of the unit can be thought of as three V-shaped planar wires, welded together at their ends. There is a sheet of the foam surface hanging on each wire, and these sheets meet along a curved triple junction.

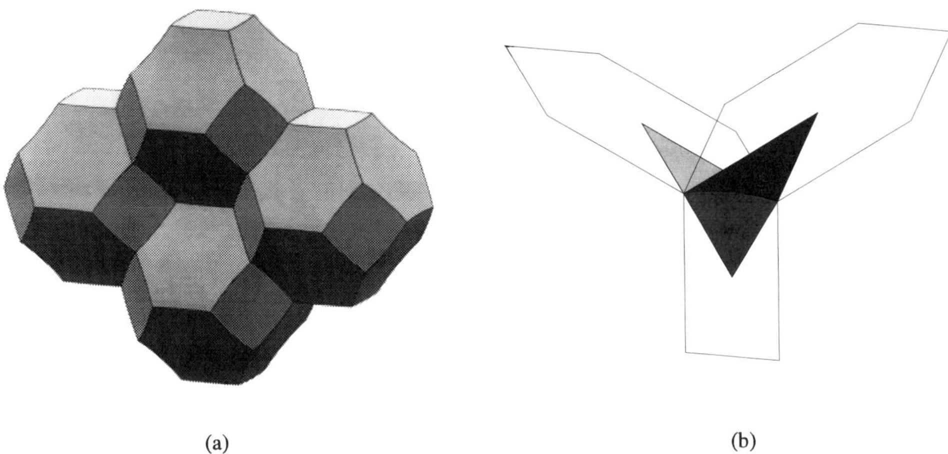

(a) (b)

Fig. 3. Four cells in the relaxed Kelvin foam (a), and one fundamental piece (b), consisting of a fourth of one square and sixths of two hexagons, the region near a particular triple edge. The outlined polygons in (b) show the original locations of nearby triple edges to help locate the figure; of course in the relaxed foam these would also bend.

We can give a lower bound for the area of this (or any partition in the Kelvin pattern) by a *slicing argument*. We will slice the fundamental unit just described by planes perpendicular to the original edge. Each slicing plane meets the three boundary wires of the unit, and the slice of the surface is Y-shaped, connecting these three points. It thus has at least as much length as the Steiner tree on these points. Furthermore, the area of the whole unit is at least the integral of the lengths of the slices. In fact, because the boundary wires are linear, the Steiner trees in the parallel planes are similar to one another, so it is easy to compute this lower bound explicitly (ALMGREN *et al.*, in preparation). We find that the cost of the Kelvin foam is bounded below by $\mu > (27/16)(\sqrt{3/2} + 1)^3 > 18.5816$. (Experiments with the Evolver show partitions in the Kelvin pattern with cost just under 18.67582; presumably this is close to the true cost of the Kelvin foam.)

3. The Weaire-Phelan Foam

Certain metals, like the β-form of tungsten or a chromium-silicon alloy, crystallize with atoms in a close packing called A15. This includes the sites of the BCC lattice together with half of its Voronoi corners: the points cyc(0,±1,±2) and their translates by CUB. Weaire and Phelan based their foam on the pattern of the Voronoi cells for this A15 packing. The pattern has a (pentagonal) dodecahedron centered at each BCC lattice site, and a certain 14-hedron around each of the other sites, as seen in Fig. 4. This 14-hedron might be viewed

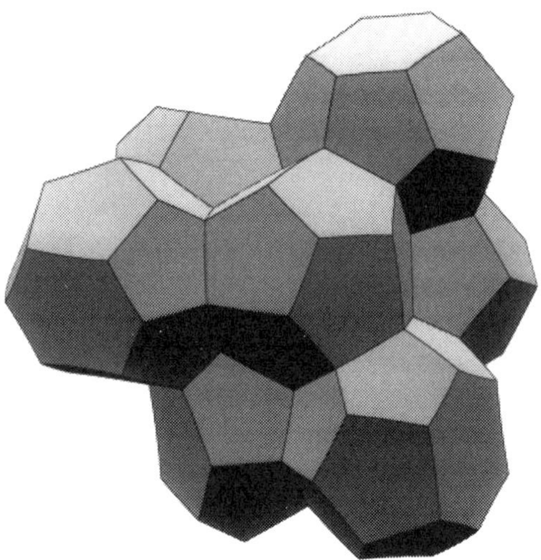

Fig. 4. Eight cells, forming a fundamental unit of the Weaire-Phelan foam. In front, we see a dodecahedron with slightly higher pressure than the neighboring cells. These are 14-hedra, with two parallel hexagonal faces, stacked in columns in the three coordinate directions.

as a "(6/5)-fold unwrapping" of a pentagonal dodecahedron, since it has two opposing hexagon faces and two belts of six pentagons surrounding these hexagons.

The symmetries of the pattern (crystallographic group $Pm\overline{3}n$) form a subgroup G of index two in G_0. This subgroup is generated by the CUB lattice translations, cyclic permutations and sign changes of the coordinates (geometrically, three-fold rotation and mirror symmetries), together with a motion which combines translation by (2,2,2) with an odd permutation of the coordinates (geometrically, translation and an order-four rotation about a coordinate axis).

In one translational unit cell for the CUB lattice $4\mathbb{Z}^3$ (whose volume is $4^3 = 64$) we have two dodecahedra and six 14-hedra. For an equal-volume partition, each cell should have volume $V = 8$. Up to symmetry in the pattern, there are two kinds of cells, three kinds of faces (hexagons, pentagons on dodecahedra, and other pentagons), four kinds of edges, and three kinds of vertices (those touching zero, one, or two hexagons).

Any dodecahedron around the origin with this symmetry will have eight corners at $(\pm a,\pm a,\pm a)$, which are the vertices touching no hexagons, and twelve corners $\mathrm{cyc}(0,\pm b,\pm c)$, the ones touching one hexagon. We take $0 < b < a < c < 2$ to fix orientation. (Choosing instead $b > c$ would give the pattern rotated by 90°, or translated by (2,2,2).) These parameters are not fixed by the symmetry. We will not try to describe the actual least-area foam in this pattern, but only an equal-volume polyhedral partition, the appropriately weighted Voronoi partition.

The pentagonal faces are planar if $(a-c)^2 = a(a-b)$. Then their area is $(a\sqrt{a-b} + (a+b)\sqrt{a})\sqrt{2a-b}$. The dodecahedron's volume is $4a(2ac + bc - ab)$.

To examine the other cells, write an arbitrary point $(1+p, 0+q, 2+r)$ as $[p,q,r]$. Then the cell near $(1,0,2) = [0,0,0]$ has dodecahedral neighbors around $[-1,0,\pm 2]$, $[1,\pm 2,0]$, and other neighbors around $[\pm 2,0,0]$, and $[1,\pm 1,\pm 2]$ and $[-1,\pm 2,\pm 1]$. Since we know the vertices of the dodecahedra in terms of a, b, c, we can list the vertices this cell shares with its dodecahedral neighbors:

$$[1-a,\pm(2-a),\pm a],\ [1-c,\pm(2-b),0],\ [1,\pm(2-c),\pm b],$$

and

$$[a-1,\pm a,\pm(2-a)],\ [c-1,0,\pm(2-b)],\ [-1,\pm b,\pm(2-c)].$$

The remaining vertices (those touching two hexagons, and four of the 14-hedra), are actually fixed by the symmetries: the vertex between the cells around $[0,0,0]$, $[2,0,0]$, $[1,1,2]$, and $[1,-1,2]$ is at their center of mass $[1,0,1]$, so the vertices of our cell not listed above are $[1,0,\pm 1]$ and $[-1,\pm 1,0]$.

We now compute the areas of the faces of this cell not shared with dodecahedra. The face to the cell at $[2,0,0]$ is a planar hexagon, with area $4(2-c)b + 2(2-c)(1-b) = 2(2-c)(1+b)$. The face to the cell at $[1,1,2]$ is a pentagon with vertices

$$[1,0,1],\ [1,2-c,b],\ [1-a,2-a,a],\ [-1+a,a,2-a],\ [-1+c,0,2-b].$$

This will be planar if $2b = c$. Making this assumption, we can divide it into one isosceles triangle of area $(1-a)\sqrt{6}$ and two congruent triangles of total area $a(1-b)\sqrt{6}$.

The foam within a unit cell of CUB has two dodecahedra and six other cells, and thus has 24 dodecahedral faces, six hexagonal faces, and 24 of the other pentagonal faces. The area of these taken together is $8A$ where A is the average area per cell.

Under our assumptions of planarity, $4b = 2c = 3a$, and it is easy to check that each face is perpendicular to the line connecting the cell centers, so that our polyhedral foam is actually a weighted Voronoi partition on these centers, where the relative weighting between the BCC centers and other centers depends on c. Also, we compute that the area is $8A = 24 + 24\sqrt{6} + (12\sqrt{5} - 8\sqrt{6} - 6)c^2$, while the volume of the dodecahedron is $V_d = 4c^3$. To get an equal volume foam, we must take $c = \sqrt[3]{2}$, which gives $A = 3 + 3\sqrt{6} + (6\sqrt{5} - 4\sqrt{6} - 3)/\sqrt[3]{16}$ and $V = 8$. The cost ratio $\mu = A^3/V^2$ for this partition is then just under 18.57752.

We have thus shown that even this unrelaxed, polyhedral partition in the A15 pattern has less cost than any Kelvin partition. Experiments with the Evolver give better partitions in this pattern, with cost just under 18.4871. Presumably there is a unique Weaire-Phelan foam in this pattern, with cost close to this value. The lower symmetry of the A15 pattern compared to the Kelvin pattern, however, leaves mathematical existence results and cost bounds out of reach.

4. Other Possible Foam Structures

Because the edges of any foam meet at tetrahedral angles arccos(−1/3) ≈ 109.47°, which is close to the angle of a regular pentagon, there has been repeated speculation (COXETER, 1958; WILLIAMS, 1968; RIVIER, 1994) that efficient foams would have pentagonal faces, unlike Kelvin's foam. Of course, using only pentagons, each cell would be a dodecahedron, and these do not tile Euclidean space. However, there is a large class of foams in which the faces are all pentagons or hexagons, as in the Weaire-Phelan foam.

There are only four types of polyhedra which have only pentagonal and hexagonal faces, with no adjacent hexagons. These are the dodecahedron, the 14-hedron we have already seen,-and two more polyhedra, with 15 and 16 faces, again each including twelve pentagons. Chemists have observed many transition-metal alloys in which each atom has a Voronoi cell of one of these types (and thus, valence 12, 14, 15, or 16). These crystals are called *tetrahedrally close-packed* (TCP) structures (FRANK and KASPER, 1958). The same structures are also observed chemically in clathrates, like chlorine hydrates, with oxygen atoms at the Voronoi corners bonded by hydrogen along the Voronoi edges, forming cages in which large gas molecules sit (at the sites of the metal atoms in the first structure).

There are several dozen TCP structures observed by the chemists, and further infinite families can be constructed by mixing these: in fact, there seem to be three basic structures—A15, Z and C15—from which all the TCP structures can be made (see

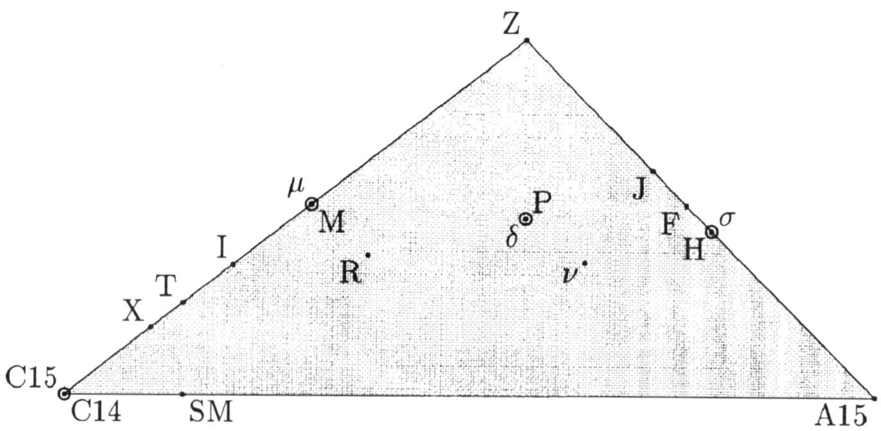

Fig. 5. Every known TCP structure, when described by its numbers of 12-, 14-, 15-, and 16-sided cells, is a convex combination of the three basic structures A15, Z, and C15. Here the horizontal axis plots the average number of faces per cell (ranging from 13(1/3) to 13(1/2)) and the vertical axis plots the fraction of cells which are 15-hedra (ranging from 0 to 2/7). (Not all known structures are labeled in this picture.) Sometimes distinct structures appear at the same point in this triangle, but they need not then have the same properties.

Fig. 5). For each, the Voronoi partition can be relaxed to an equal-volume foam as before. For Z or A15 or any mixture of these, the foam seems to have cost less than that of Kelvin, with the cost decreasing as we approach A15; the Weaire-Phelan foam obtained there seems optimal among all TCP foams (KRAYNIK *et al.*, in preparation). This gives some evidence that this Weaire-Phelan foam may be the optimal solution to Kelvin's problem.

However, many of the TCP foams, like C15, T, and SM, have a cost significantly greater than Kelvin's. The fact that the Kelvin foam (with no pentagons) has such low cost means we must reexamine our hypothesis that good foams need pentagons. In fact, we can find foams based on Voronoi cells for other crystal structures, like that of γ-brass, which do reasonably well, despite having even triangles among their faces. Thus it is hard to even guess where else to look for potential competitors for the optimal equal-volume foam, and we see little hope for any rigorous solution to Kelvin's problem; even the analogous problem in the plane is still open.

The fact that the Kelvin foam is not optimal suggests several related open problems worthy of consideration (SULLIVAN and MORGAN, 1996). It is interesting to note that the Weaire-Phelan foam does not have *equal pressures* in its cells, while the Kelvin foam does. (There has been repeated confusion in the literature between the equal-pressure and equal-volume conditions.) In fact, it seems that no foam in the A15 pattern (even allowing differing volumes) can have equal pressures.

In 1986, Kusner derived some necessary conditions for equal-pressure foams, such as a lower bound of $2 + 2\pi/(3 \arccos(1/3) - \pi) > 13.39733$ on the average number of faces for a cell in the foam (KUSNER, 1992). Choe proved that every compact three-manifold has a least-boundary-area fundamental domain (CHOE, 1989); the *Choe cells* for Euclidean manifolds all give equal-pressure, equal-volume foams in space. In fact, we know of no other equal-pressure foams. Perhaps these are the only combinatorial types of cells which can occur in such a foam. Presumably, for instance, no TCP pattern could yield an equal-pressure foam; many are already ruled out by the lower bound above. We conjecture that the Kelvin foam is the Choe cell for the BCC torus, and is the best foam which has equal pressures as well as equal volumes.

REFERENCES

ALMGREN, F. (1991) Computing soap films and crystals, in *Computing Optimal Geometries*, Selected Lectures in Mathematics, pages 3–5 plus video, Amer. Math. Soc.

ALMGREN, F., KUSNER, R. and SULLIVAN, J. M., A comparison of the Kelvin and Weaire-Phelan foams (in preparation).

ANGENENT, S., CHOPP, D. and ILMANEN T. (1995) Nonuniqueness of mean curvature flow in $\mathbb{R}^3$, *Comm. Part. Diff. Eq.*, **20**, 1937–1958.

BRAKKE, K. A. (1992) The Surface Evolver, *Exper. Math.*, **1**, 141–165.

CHOE, J. (1989) On the existence and regularity of fundamental domains with least boundary area, *J. Diff. Geom.*, **29**, 623–663.

COXETER, H. S. M. (1958) Close-packing and froth, *Ill. J. Math.*, **2**, 746–758.

FRANK, F. C. and KASPER, J. S. (1958) Complex alloy structures regarded as sphere packings. I. Definitions and basic principles, *Acta Crystallographica*, **11**, 184–190.

KRAYNIK, A. M., KUSNER, R., PHELAN, R. and SULLIVAN, J. M., TCP structures as equal-volume foams (in preparation).

KUSNER, R. (1992) The number of faces in a minimal foam, *Proc. R. Soc. Lond.*, **439**, 683–686.

OKABE, A., BOOTS, B. and SUGIHARA, K. (1992) *Spatial Tessellations: Concepts and Applications of Voronoi Diagrams*, Wiley & Sons.

PLATEAU, J. (1873) *Statique Expérimentale et Théorique des Liquides Soumis aux Seules Forces Moléculaires*, Gauthier-Villars, Paris.

RIVIER, N. (1994) Kelvin's conjecture on minimal froths and the counter-example of Weaire and Phelan, *Europhys. Lett.*, **7**, 523–528.

SENECHAL, M. (1990) *Crystalline Symmetries*, Adam Hilger.

Sir William THOMSON (Lord KELVIN) (1887) On the division of space with minimum partitional area, *Phil Mag.*, **24**, 503–614.

SULLIVAN, J. M., The VCS software for computing Voronoi diagrams, available by email from sullivan@geom.umn.edu.

SULLIVAN, J. M. and MORGAN, F. (editors) (1996) Open problems in soap bubble geometry: Posed at the Burlington Mathfest in August 1995, *Int'l J. Math.*, **7**, 833–42.

TAYLOR, J. E. (1976) The structure of singularities in soap-bubble-like and soap-film-like minimal surfaces, *Ann. of Math.*, **103**, 489–539.

WEAIRE, D. and PHELAN, R. (1994) A counter-example to Kelvin's conjecture on minimal surfaces, *Phil. Mag. Lett.*, **69**, 107–110.

WILLIAMS, R. E. (1968) Space filling polyhedron: Its relation to aggregates of soap bubbles, plant cells, and metal crystallites, *Science*, **161**, July, 276–277.

Kelvin's 'Division of Space', and Its Aftermath

D. A. ABOAV

29 Clements Road, Chorleywood, Hertfordshire, WD3 5JS, U.K.

Keywords: Foam, Honeycomb, Origin of Life

1. Introduction

In 1887 Sir William THOMSON (Lord KELVIN) submitted to the *Philosophical Magazine* a paper entitled "On the division of space with minimum partitional area", opening with the words "This problem is solved in foam". The object of that work—the Kelvin Problem—was to construct a three-dimensional cellular structure, or honeycomb, of 'equal and similar' polyhedral cells 'sameways oriented', with faces of minimal area. Kelvin attacked the problem in two stages: first he made a honeycomb of that type out of straight-edged, plane-faced polyhedra, whose shape was predetermined by the requirement that the cells be not only 'space-filling', but also congruent and similarly oriented. With these metrical restrictions it was inevitable that an individual cell of the desired honeycomb would not have the smallest surface area possible for a polyhedron of its volume and number of faces. That is because a trihedral, N-hedron required to fill space is excluded on topological grounds from being the N-hedron with the smallest surface area for its volume; indeed, as we shall see, the space-filling requirement alone obliges the surface area of such a polyhedron of given volume to be nearly 2% greater than the minimum compatible with its number of faces (GOLDBERG, 1934). Only after he had fixed the shape of the cell in that way did Kelvin, in a second operation, proceed to make the area of the individual faces as small as possible, by curving them slightly; but this reduced their area, and hence the partitional area of the honeycomb, by a mere 0.2% or so (PRINCEN and LEVINSON, 1987), that is, by considerably less than the amount by which he had had, by the initial restriction, to make the surface area of individual cells exceed the minimum possible. It is worth keeping this fact in mind, lest we should later seek to make the partitional area of such a honeycomb smaller still, by subjecting it to conditions less severe than Kelvin's.

Kelvin thus showed the cell of the desired honeycomb to be the semi-regular, truncated octahedron of Archimedes, but with its hexagonal faces so curved as to make the faces and edges of the cells meet at angles of arccos $-1/2$ (120°) and arccos $-1/3$ (109°28′16″23‴ ...), respectively. He was later (1894) to call the plane-faced polyhedron 'orthic', and the equiangular one, 'orthoidal'.

2. The Isoperimetric Problem for Polyhedra

In Figs. 1(a) and 1(b) are shown two trihedral 14-hedra of equal volume, with equal edges, and with faces that are plane and hence of minimal area. They are: (a) a semi-regular, truncated octahedron—i.e. Kelvin's orthic polyhedron—and (b) a truncated, hexagonal, double skew pyramid. The number shown beneath each polyhedron is the ratio of the surface area of the polyhedron to that of a sphere of equal volume, and is about 2% smaller for the second figure than for the first. This shows there is at least one 14-hedron with a surface area smaller than that of an orthic Kelvin polyhedron of the same volume.

Despite its elementary nature, the problem of determining the polyhedron with the smallest surface area for a given volume and given number of faces is not easy to solve. Without entering on the matter in any detail we may notice the following:

(i) A necessary condition is that the faces of the polyhedron be tangent to a sphere at the centre of gravity of the faces (LINDELÖF, 1869); and this condition is sufficient for a polyhedron of given type, if all its vertices are trihedral (STEINITZ, 1922);

(ii) It is conjectured (GOLDBERG, 1934) that, among the N-hedra of given type, the one with the smallest surface area for a given volume is a 'medial' one—i.e. that it is a trihedral polyhedron possessing, at most, only n-gonal and $(n + 1)$-gonal faces;

(iii) For $N < 16$ there is exactly one medial polyhedron for every value of N, except $N = 11$ and $N = 13$. Consequently a trihedral 13-hedron cannot possess pentagonal and hexagonal faces only, but must have at least one of its faces of a different shape;

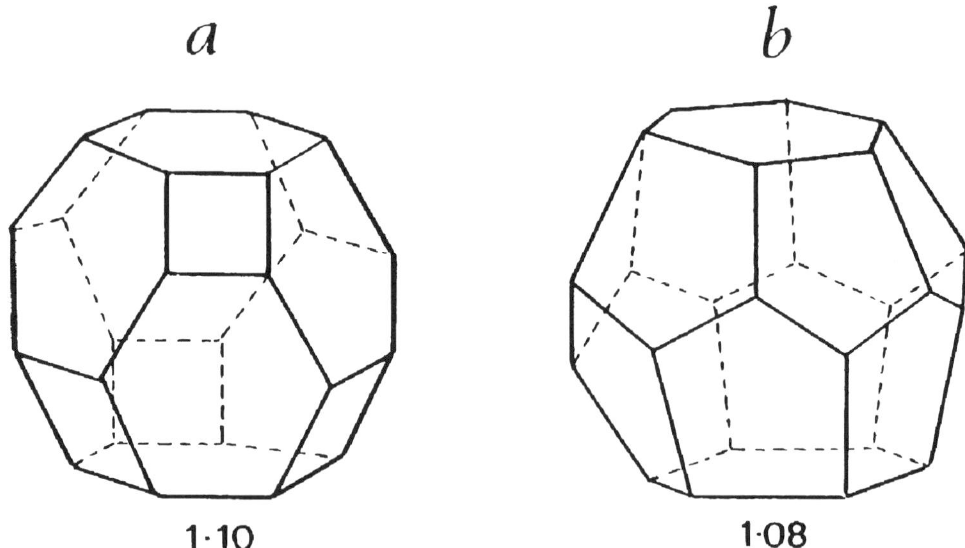

Fig. 1. (a) Truncated octahedron; (b) truncated, hexagonal, double-skew pyramid.

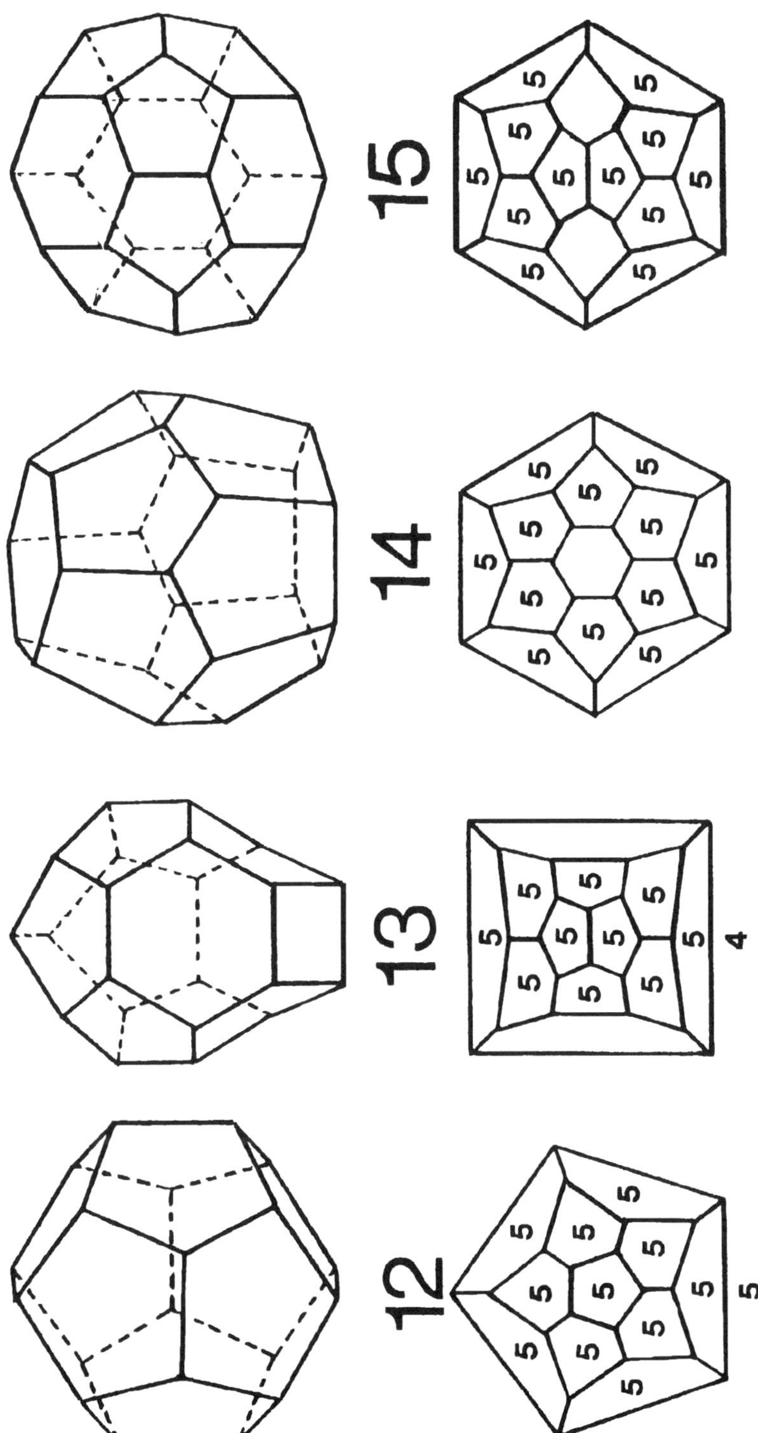

Fig. 2. First 4 polyhedra of 'augmented 12-hedral' series, and their Schlegel diagrams.

(iv) Of the many different trihedral 14-hedra only one, the truncated double skew pyramid of Fig. 1(b) is medial; so that, if Goldberg's conjecture is correct, this polyhedron is the 14-hedron with the smallest surface area for its volume.

Apart, then, for a gap at $N = 13$, the medial polyhedra with $N \geq 12$ thus form an unbroken series, the so called '12-hedral' series, the faces of whose members are pentagonal, or pentagonal and hexagonal, only; and these polyhedra are the ones with the smallest surface area (GOLDBERG, 1934). The gap in the series will here be filled by adding to it, from among the possible 13-hedra, the one with a single face quadrilateral and the remaining ones pentagonal and hexagonal. The resulting series will here be called the 'augmented 12-hedral' series. Figure 2 shows the first 4 polyhedra of this series, with their Schlegel diagrams.

The above metrical considerations suggest the type of polyhedron to be used in constructing a 4-connected honeycomb of minimal partitional area; but, to ascertain the optimal average number of faces of the polyhedra the topology of the honeycomb must be considered in further detail.

3. The 4-Connected Honeycomb

In a 4-connected honeycomb four cell-edges meet at each vertex. The angle, Θ, between each pair of them is called a *vertex angle*. There are at each vertex six such angles. They are in general unequal, and their sum is not constant. If they are equal, they have the value arccos $-1/3$, or 109°28′16″23‴ ... . This angle is not commensurable with a rightangle (Euclid, X, def. 1). In practice $\langle\Theta\rangle$, the average value of the vertex angle, generally differs from arccos $-1/3$ by less than a degree, and is related to $\langle N\rangle$, the average number of faces per cell, by the equation:

$$\langle\Theta\rangle = 120°\left(\langle N\rangle - 3\right) / \left(\langle N\rangle - 2\right). \tag{1}$$

A *regular* 4-connected honeycomb—in COXETER's (1969) notation a $\{p,3,3\}$ honeycomb—is an imaginary figure if only because its required vertex angle, in this case arccos $-1/3$, is incommensurable with a rightangle. Since all its vertex angles are assumed equal, however, such a figure may be regarded as periodic. In this imaginary honeycomb the average number $\mathcal{N}$ of faces per cell is determined by substituting for $\langle\Theta\rangle$ in Eq. (1) the value arccos $-1/3$. The resulting equation may be written:

$$\left(\langle N\rangle - 2\right)\text{arccos}-1/3 = \left(\langle N\rangle - 3\right)\text{arccos}-1/2. \tag{2}$$

$\mathcal{N}$, an irrational number, can be expressed as a non-recurring decimal 13.397 332 57 ..., or as an infinite simple continued fraction:

$$\mathcal{N} = 13 + \frac{1}{2+}\,\frac{1}{1+}\,\frac{1}{1+}\,\frac{1}{14+}\,\frac{1}{2+}\,\frac{1}{1+}\,\frac{1}{1+}\,\frac{1}{8+}\cdots. \tag{3}$$

Table 1. Convergents of $\mathcal{N}$.

r	N_r		$(\Delta\Theta)_r$
1	13(1/2)	13.5	5′ 38″
2	13(1/3)	$13.\dot{3}$	3′ 34″
3	13(2/5)	13.4	0′ 9″
4	13(29/73)	$13.\dot{3}97\ 260\ 2\dot{7}$	0′ 0″ 14‴

The first 4 values of its r-th convergent, N_r, are listed in Table 1 as vulgar (col. 2) and decimal (col. 3) fractions, together with the amount $(\Delta\Theta)_r$ by which the average vertex angle of a honeycomb for which $\langle N \rangle = N_r$ exceeds, or falls short of arccos $-1/3$.

The table lists ever-closer approximations, N_r, to $\mathcal{N}$; so that $\langle N \rangle$ by equating it to N_r can be made to approach the irrational number as nearly as we please. But it should not be ignored that $\mathcal{N}$ may also be approximated—for our purpose perhaps more conveniently—by a series of rational fractions that approach it irregularly, rather than systematically in a limiting process as do the convergents of a continued fraction; for Newton's concept of 'limit' (*Principia*, I, 1, lem. 1), like Euclid's of a 'curved' line (*Elements*, I, def. 15), is foreign to the notion of 'integer', or whole number (*ibid.* VII, def. 2), which the above considerations show to be better suited to our needs. Limitations of space do not allow this distinction, a matter of the first importance, to be enlarged upon here; but it is hoped to take it up in a later communication.

What few empirical results there are show the global, or partitional area of a 4-connected honeycomb with plane faces to be larger, the more $\langle N \rangle$ departs from $\mathcal{N}$ (RIVIER, 1994). In Kelvin's honeycomb, for example, the surface area of a 14-faced cell is, as we have seen, about 10% greater than that of a sphere of the same volume; whereas in a random Voronoi honeycomb, where the approximate average number of faces per cell is as large as 15(1/2), the average surface area of a cell is as much as 24% greater than that of a sphere (THORVALDSEN, 1992). It is therefore reasonable to suppose the partitional area of a 4-connected honeycomb to be in general smaller, the closer the average number $\langle N \rangle$ of the faces of its cells is to $\mathcal{N}$.

It seems, then, that for a periodic, 4-connected honeycomb to divide space with as small a partitional area as possible it should have a unit cell (i) which comprises more than one kind of polyhedron, chosen from among those of the 'augmented 12-hedral' series, and (ii) in which the average number of its faces lies as close as possible to $\mathcal{N}$. The possible shapes such a cell may have are suggested by the crystal structure of a particular class of chemical compound, the inclusion hydrate.

4. Inclusion Hydrates

Aqueous solutions of many gases on freezing form crystalline hydrates, whose lattice structure is that of a periodic, 4-connected honeycomb with atoms of oxygen at its vertices forming 'cages' in which molecules of gas are entrapped. Numerous such hydrates—or 'inclusion' hydrates—have been prepared and their crystal structure determined by X-ray analysis; but we shall here confine our attention to two commonly occurring types, exemplified by the hydrates of chlorine and chloroform (von STACKELBERG and MÜLLER, 1951), and here referred to as Types I and II, respectively.

The unit cell of chlorine hydrate (Type I) is made up of two 12-hedra and six of the 14-faced truncated, hexagonal bipyramids of Fig. 1(b). The average number $<N>$ of faces per cell is therefore 13(1/2). That of the hydrate of chloroform (Type II) on the other hand has eight 12-hedra and 16 truncated bipyramids, and so has $<N>$ equal to 13(1/3). These unit cells, whose constituent cells are not necessarily of the same volume, 'fill space' to form periodic, 4-connected honeycombs, of which, as the above figures show, the unit cells of Types I and II have $<N>$ equal to N_1 and N_2, that is, to the 1st and 2nd convergents of $\mathcal{N}$ (Table 1), respectively. Pen-and-ink drawings of these honeycombs, each of which has a cubic lattice structure, have been drawn by WILLIAMS (1979).

In 1994 WEAIRE and PHELAN, using a method of BRAKKE (1992) altered the metrical properties of each of these two types of microscopic honeycomb, so as to make their cells equal in volume and of minimal partitional area and found that, while the partitional area of the altered honeycomb of Type II ($<N> = 13(1/3)$) is 0.4% greater than that of the corresponding Kelvin honeycomb, that of Type I ($<N> = 13(1/2)$) is 0.3% smaller. They were thus able to demonstrate, for the first time since its inception in 1887, that Kelvin's honeycomb does not possess the smallest partitional area for a periodic, 4-connected honeycomb of a given volume and given number of cells.

This raises the question whether the partitional area of such a honeycomb cannot be made smaller still. Table 1 shows that a honeycomb with $<N> = N_4$, or 13(29/73), has an average vertex angle that differs from arccos $-1/3$ by less than a quarter of a second of arc—by less, that is, than the angle subtended at the Earth by one of Jupiter's moons—which suggests that this honeycomb, if it could be made, might be modified by the method of Brakke to acquire an exceptionally small partitional area. As a first step in that direction we here give details of a recent, unsuccessful attempt to construct it, and in conclusion consider the light its result may throw on a matter of biological interest.

5. A Periodic, 4-Connected Honeycomb with $<N> = 13(29/73)$

The problem of assembling a honeycomb of this complexity is mainly geometrical; for while, as the following example shows, it is easy to choose a numerical distribution of differently shaped polyhedral cells to give a symmetrical, composite cell with the required average number of faces per polyhedron, it is difficult to determine such a distribution that allows the cells to 'fill space', and so to serve as the 'unit cell' of a honeycomb.

Plate 1, for example, shows a composite cell of 146 polyhedra of the augmented 12-hedral series, which for brevity will here be referred to as the 146-cell. It is of hexagonal symmetry and has in all 1956 faces, distributed as follows:

Plate 1. '146-cell' of postulated *zoöhydrate*, or 'ice of life'.

	2	symmetrical 12-hedra	with a total of	24	faces (grey)
	11	left-handed 12-hedra	with a total of	132	faces (light green)
	11	right-handed 12-hedra	with a total of	132	faces (dark green)
	24	left-handed 13-hedra	with a total of	312	faces (light red)
	24	right-handed 13-hedra	with a total of	312	faces (dark red)
	36	left-handed 14-hedra	with a total of	504	faces (light blue)
	36	right-handed 14-hedra	with a total of	504	faces (dark blue)
	1	left-handed 18-hedron	with	18	faces (yellow)
	1	right-handed 18-hedron	with	18	faces (amber)
i.e.	146	polyhedra	with a total of	1956	

The average number of faces per polyhedron is therefore 1956/146, or 13(29/73) as required. It is found that, if they are to join together without gaps, the 12-, 13- and 14-hedra, with the exception of the 12-hedra lying on the a-axis and the two 18-hedra, must be asymmetrically distorted, to form equal numbers of left- and right-handed geometric isomers. The two 18-hedra, of which one is at the centre of the cell and so hidden from view, are enantiomorphs: one left- and the other right-handed. They are coloured yellow and amber, respectively; and their Schlegel diagrams are shown in Fig. 3.

The 146-cell has the shape of a dumb-bell, with a left-handed and a right-handed half, each containing 73 polyhedra and each the mirror image of the other. To see how such a cell may join with others like it to form a multicellular aggregate, it is convenient first to examine briefly its construction.

The 144 polyhedra that remain if the two 18-hedra are removed can be divided into two groups of 72, each of which has twelve 12-hedra, twenty-four 13-hedra, and thirty-six 14-hedra symmetrically arranged about the a-axis, and has roughly the shape of an

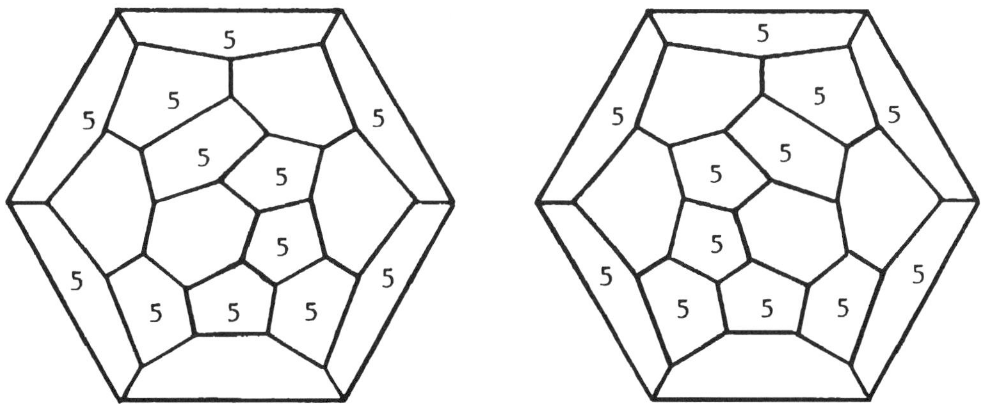

Fig. 3. Schlegel diagrams of enantiomorphic 18-hedra, left-handed and right-handed.

oblate spheroid with the a-axis as its minor axis (Plate 1). Since there is a single 12-hedron at the centre of the group, and since the polyhedra are 72 in number, one of the outside 12-hedra must be left out of account if the cell is shown on its own; for two of these 12-hedra are 'shared' by adjacent cells as they join along their c-axis (see below).

The polyhedra at the upper end of a 72-cell fit exactly on to those at the lower end of its inversion, rotated about the a-axis through 30°, and so enable such cells, alternately inverted, to line up in a row along their a-axis. The 'gaps' between pairs of 72-cells brought together in that way have the shape of the enantiomorphic 18-hedra of Fig. 3, alternately left- and right-handed: into them the removed 18-hedra (one left-, one right-handed), can therefore be fitted to re-form the 146-cell with its two halves, one left-handed and the other right-handed. By linking the cells in that way, the first stage in the construction of a honeycomb, namely the lining-up of the 146-cells along their a-axis, is achieved.

The joining of cells along their c-axis is not so straightforward, and is difficult to visualize without the help of a model like that of Plate 1. Such a model, which it was unfortunately not possible to have ready in time to illustrate this article, will therefore be described in outline only, as follows:

A pair of 72-cells with parallel a-axes can be joined along a c-axis, if rotated relatively to each other through 120 ° about that axis. So joined they allow a left-handed (or right-handed) 18-hedron to be added, at only one end of the a-axis of each cell, to form a left-handed (or right-handed) 73-cell, according as the rotation is clockwise (or anti-clockwise).

The addition of further 72-cells in that way produces a left-handed (or right-handed) 'helix', whose cells go through a complete (360°) turn at every third step.

To each 72-cell of the helix can be added, in the same way as before, a left-handed (or right-handed) 18-hedron, according as the helix has the form of a left-handed (or right-handed) screw.

The joining of 72-cells along a c-axis thus differs from that along the a-axis in that:

(a) the resulting chain, or 'filament' of 73-cells is twisted like a left-handed (or right-handed) helix; and

(b) the 18-hedra are situated at the outside of the filament, instead of on its axis.

The above description, though no more than a sketch, suffices to show that a periodic, space-filling honeycomb cannot be made by joining 146-cells along both their a- and their c-axes, since this gives rise to a structure comprising additional, irregular spaces, which cannot be eliminated without packing the polyhedra more closely, and so destroying the cells' symmetry. One such closer packing is effected by a *polymorphic transformation* in which the structure is split into component helices of 73-cells, the 'open' honeycomb thereby breaking up into equal numbers of independent, isomeric, left-handed and right-handed helical filaments, which occupy a smaller space without, however, filling it. It is therefore evident that of the composite cells with $\langle N \rangle = 13(29/73)$ this particular one will not help solve the Kelvin Problem; and consequently that, among the periodic, 4-connected honeycombs so far reported, that of Weaire and Phelan is still the one with the smallest partitional area.

6. An Inclusion Hydrate with <*N*> = 13(29/73)

We bring to a close our discussion of Kelvin's 'Division of Space' by noting that the fact that the 146-cell will not form a honeycomb without gaps does not exclude that there may exist an inclusion hydrate whose molecular structure can be represented by such a cell; for (a) its polyhedra are of the required type, and (b) molecules of matter in the crystalline state do not always 'fill space': their spatial arrangement may for example "lead to some kind of structure which, although regulated by definite 'rules', cannot extend indefinitely for purely topological and geometrical reasons" (WELLS, 1970) and, if no alternative structure is available, may exist indefinitely in the form of a system of polyhedra "that does not form part of any possible three-dimensional honeycomb" (*ibid.*). One is thus led to consider the possible chemical composition of an inclusion hydrate having the structure of a 146-cell, and the circumstances that might bring about its synthesis.

A clue as to what compound might be 'trapped' in such a hydrate is given by the helical structure resulting from the above-mentioned polymorphic transformation. Since such a structure is a characteristic of a polypeptide (PAULING *et al.*, 1951) one's first thought is that there may exist an inclusion hydrate which has the structure of the 146-cell and which encloses one or more amino-acids, or related compounds, even though (JEFFREY, 1984) such compounds are typically of the kind that render inclusion hydrates unstable.

The testing of such a supposition is facilitated by the fact that the structure of the 146-cell, and hence the pattern of its corresponding Laue photograph, are known with precision before ever a synthesis of the sought-after hydrate is contemplated. If, then, such a compound exists, it should not take long to identify, since what is generally the most laborious part of such an undertaking, namely the determination of the crystal structure, is in this case unnecessary. From a number of such hydrates prepared for the purpose the sought-after one, if it happens to be among them, will be immediately recognizable from its Laue photograph; and, if it is not, more can be prepared until eventually one is found with the desired structure.

Let us suppose we succeed in this way in identifying such a hydrate with the structure of an assembly of 146-cells. The question will then arise whether the same compound may not also exist in the more closely-packed form of helical filaments, either left-handed or right-handed.

Since inclusion hydrates, especially porous ones, are readily destabilized under pressure to give new hydrates (DYADIN *et al.*, 1991), it is to be expected that, under pressure, an inclusion hydrate having the 'open' structure of an assembly of 146-cells might readily undergo the polymorphic transformation needed to produce its more closely packed 'helical' isomer. The first task on identifying a hydrate of that structure would therefore be to subject it to pressure at a sufficiently low temperature and to observe whether at any stage a sudden change occurs in its Laue pattern. If such a change were in fact observed, and if the resulting pattern were to turn out to be that of a helical structure, a simple means might thereby have been found to synthesize a polypeptide-like compound from its elementary components. Such a supposition might understandably be dismissed as idle conjecture, were it not for widespread evidence to suggest that a reaction of that kind may have occurred naturally at an early stage in our planet's history.

It is generally acknowledged that the Earth's atmosphere, as well as containing water vapour as at present, once resembled the present-day atmosphere of Jupiter in that it had methane and ammonia among its main constituents (UREY, 1952); and it has further been demonstrated that, in the presence of an electrical discharge like a flash of lightning, these three hydrides (CH_4, NH_3, OH_2) react to produce, among other substances, traces of various amino-acids (MILLER and ORGEL, 1974). As the Earth cooled, the first of the hydrides to liquefy would be that of oxygen, which would condense on its surface to form lakes and oceans of water, with which the various constituents of the atmosphere, including those formed through the action of lightning, would be in chemical equilibrium. As the Earth cooled still further, frozen water falling as hail and snow would eventually settle in colder regions like the poles and mountain peaks to form glaciers, a condition in which chemical compounds dissolved in it tend to form hydrates. As such glaciers melted, dissolved or trapped compounds, if they had undergone no further chemical change within the ice, would be released to rejoin the liquid-gas equilibrium of ocean and atmosphere. These putative chemical reactions are based on facts and experiment, and invoke no special hypothesis.

Where a special hypothesis is called for, however, is in accounting for the fate of an amino-acid hydrate having the structure of the model illustrated in Plate 1, if it happens to be trapped within a glacier at a sufficiently great depth below the surface.

If the glacier is deep enough, the pressure exerted at its base by the superincumbent ice may suffice to bring about a transition of a kind envisaged by DYADIN *et al.* (*v. supra*), such as might convert the hydrate of one or more amino-acids having the above structure into its more closely packed polypeptide-like isomer. If so, it does not follow that the glacier would on melting give up the resulting compound to the atmosphere, since the activation energy needed to restore it to the condition required for that to occur might no longer be available (WELLS, 1984), and the compound might therefore dissolve in the water of the ocean or alpine lake, its molecules retaining their newly acquired, stranded, helical structure. In this connection it should be remembered that at so early a stage in the planet's history there would have been no living organisms present to consume the new compound as it was formed, and thereby to prevent its further chemical evolution.

One cannot fail to be struck by the resemblance of the helical structure of this supposed compound to that of one of the basic constituents of living organisms; and, although it cannot at present be demonstrated whether, or how those organisms could have evolved from a compound of the above composition and structure, there is at least a presumption that they may have done so. Life on Earth may then have originated, not as Darwin surmised in 'some warm little pond' (DARWIN, 1871), but in the depths of a massive, cold glacier.

It is, then, not unreasonable to suppose that there exists a hydrate of an amino acid, or related compound, with the structure of the composite cell illustrated in Plate 1; that, over long periods of time in the distant past, this compound was continually synthesized in and precipitated from the Earth's atmosphere; and that the typically helical molecular structure associated with the living forms now covering the surface of the Earth evolved from it. It may therefore be appropriate to call this postulated hydrate the *zoöhydrate*, or 'ice of life'.

Since inclusion hydrates are not difficult to prepare in the laboratory, the question whether there exists one of the above composition and structure is easily settled by experiment; and to speculate further on its likely existence is therefore pointless.

REFERENCES

BRAKKE, K. (1992) The Surface Evolver, *Exp. Math.*, **2**, 141–165:

COXETER, H. S. M. (1969) *Introduction to Geometry*, 2nd Ed., John Wiley & Sons, Inc., NY, pp. 411–442.

DARWIN, C. (1871) Letter in G. de Beer, *Charles Darwin, a Scientific Biography*, Doubleday & Co., Inc., NY, **1965**, p. 271.

DYADIN, Y. A., BONDARYUK, I. V. and ZHURKO, F. V. (1991) Clathrate hydrates at high pressures, in *Inclusion Compounds 5* (eds. J. L. Attwood, J. E. D. Davies and D. D. MacNicol), Oxford University Press.

GOLDBERG, M. (1934) The isoperimetric problem for polyhedra, *Tohoku Math. J.*, **40**, 226–236.

JEFFREY, G. A. (1984) Hydrate inclusion compounds, in *Inclusion Compounds 1*, Academic Press, London.

KELVIN Lord (1894) On homogeneous division of space, *Proc. Roy. Soc.*, **55**, 1–17.

LINDELÖF, L. (1869) Propriétés générales des polyèdres etc., *St. Petersburg: Bull. Ac. Sc.*, **14**, 258–259.

MILLER, S. L. and ORGEL, L. E. (1974) *The Origins of Life on Earth*, Prentice Hall Inc., pp. 84–87.

PAULING, L., COREY, R. B. and BRANSON, H. R. (1951) The structure of proteins: two hydrogen-bonded helical configurations of the polypeptide chain, *Proc. Nat. Acad. Sci.*, **37**, 205–211.

PRINCEN, H. M. and LEVINSON, P. (1987) *J. Colloid Interface Sci.*, **120**, 172.

RIVIER, N. (1994) *Phil. Mag. Lett.*, **69**, 297–303.

VON STACKELBERG, M. and MÜLLER, H. R. (1951) *J. Chem. Phys.*, **19**, 1319.

STEINITZ, E. (1922) Polyeder und Raumeinteilungen, in *Encyk der Math Wiss*, Berlin, **III-12**, 38–43, 94–101.

THOMSON, W. (1887) On the division of space with minimum partitional area, *Phil. Mag.*, **24**, 503–514.

THORVALDSEN, A. (1992) A stochastic field approach to grain growth, in *Grain Growth in Polycrystalline Metals, Part 1* (eds. G. Abbruzzese and P. Brozzo), Trans Tech Publications Ltd., Switzerland, pp. 361–366.

UREY, H. C. (1952) On the early history of the Earth and the origin of life, *Proc. Nat. Acad. Sci.*, **38**, 351–363.

WEAIRE, D. and PHELAN, R. (1994) *Phil. Mag. Lett.*, **69**, 107–110.

WELLS, A. F. (1970) *Models in Structural Inorganic Chemistry*, Oxford University Press, p. 19; (1984) *Structural Inorganic Chemistry*, O.U.P., p. 10.

WILLIAMS, R. W. (1979) *The Geometrical Foundation of Natural Structure*, Dover Publications Inc., NY, pp. 188–189.

Elastic-Plastic Behavior of a Kelvin Foam

A. M. KRAYNIK[1] and D. A. REINELT[2]

[1]*Engineering Sciences Center, Department 9112 MS 0834, Sandia National Laboratories, Albuquerque, NM 87185-0834, U.S.A.*
[2]*Department of Mathematics, Southern Methodist University, Dallas, TX 75275-0156, U.S.A.*

Keywords: Elastic-Plastic Behavior, Foam, Kelvin Cell, Minimal Surfaces, Rheology

A child's fascination with soap bubbles carries into adulthood for physicists, mathematicians, chemists, and engineers who want to understand soap froth.

1. Introduction

Imagine the universe filled with a perfectly ordered foam that contains mostly gas. The polyhedral bubbles in this dry foam have identical volume, shape, orientation, and pressure; only their positions are different. This situation underlies the classic paper by KELVIN (1887) "on the division of space with minimum partitional area" that gave us the Kelvin cell. The Kelvin foam is a natural starting point for developing mathematical models of foam rheology in 3D just as the regular hexagonal structure was appropriate in 2D.

Kelvin's unpublished notebooks indicate that he anticipated recent theoretical studies on foam rheology in 2D a century before they appeared (D. Weaire, 1995, private communication). Kelvin used 2D as a guide to thinking about 3D, exactly as we do today. Many interesting questions concerning the microrheology of foams as complex fluids are common to both dimensions despite obvious differences between them.

PRINCEN (1983) investigated elastic-plastic behavior for simple shearing flow of perfectly ordered, fluid-fluid honeycombs that range in structure from hexagonal film networks to close-packed circles as the fraction of continuous phase increases. Subsequent investigators have extended Princen's approach to address other aspects of the 2D problem relating to cell-orientation effects (KHAN and ARMSTRONG, 1986; KRAYNIK and HANSEN, 1986), disordered structure (WEAIRE and KERMODE, 1983, 1984; WEAIRE and FU, 1988; HERDTLE, 1991), polydisperse hexagonal structure (KRAYNIK *et al.*, 1991), viscous phenomena (SCHWARTZ and PRINCEN, 1987; REINELT and KRAYNIK, 1989, 1990), and Plateau borders (BOLTON and WEAIRE, 1992; HUTZLER *et al.*, 1995). Additional references can be found in reviews by KRAYNIK (1988) and WEAIRE and FORTES (1994).

This article summarizes our current understanding of microrheological behavior associated with elastic-plastic response of a Kelvin foam. Capillary statics determines the foam structure in this regime of large quasi-static deformations; viscous effects can be neglected when the deformation rate is very small. The microrheological approach draws connections between foam geometry and forces at the cell level and the effective stress and strain of the foam at the macroscopic level. This article focuses on the geometry of deformed Kelvin cells. Mathematical details and additional results are available elsewhere (REINELT, 1993; REINELT and KRAYNIK, 1993, 1996; KRAYNIK and REINELT, 1996a, 1996b).

2. Foam Geometry

The principles that govern the geometry of a dry foam under static conditions can be traced to PLATEAU (1873). Polygonal cells in 2D are separated by circular arcs with constant mean curvature that meet at equal angles of 120°. The entire foam structure can be represented by five numbers per cell: the coordinates of two vertices and a bubble pressure. By contrast, polyhedral cells in 3D are separated by minimal surfaces with uniform mean curvature; three faces meet at equal dihedral angles along each cell edge; and four edges meet at each cell vertex at equal tetrahedral angles of $\cos^{-1}(-1/3) \approx 109.47°$. No face can be a flat polygon with straight edges. In general, both principal curvatures vary over a particular face while their sum, the mean curvature, remains constant. Closed form solutions for the shape of these minimal surfaces are unavailable, even for a Kelvin cell.

Kelvin obtained an approximate solution for the geometry of his minimal tetrakaidecahedron by solving Laplace's equation, the linearized version of the differential equation that corresponds to the Young-Laplace equation with equal bubble pressures. Finite difference solutions to the full nonlinear PDE have been used to study large extensional deformations of a Kelvin cell. However, the Surface Evolver, which was developed by BRAKKE (1992, 1995), has now become the standard software for computing the geometry of soap foams. This simulation method is based on the process of evolution by mean curvature (BRAKKE, 1977). The Surface Evolver was used to compute the minimal foam geometries in this article. In all calculations, the interfacial tension T and cell volume V are equal to one; this corresponds to scaling length by $V^{1/3}$ and stress and interfacial energy density by $T/V^{1/3}$.

The regular tetrakaidecahedron and the undeformed Kelvin cell shown in Fig. 1 have the same topology. To achieve equal dihedral angles and minimize surface area, the squares and hexagons on the former acquire curved edges and the hexagons become nonplanar. The total surface area decreases by a mere 0.16% (to 5.30627602) but other geometric features change more substantially, e.g. the dihedral angles go from 109.47° and 125.26° to 120°, the vertex angles change from 90° and 120° to 109.47°, and the area of the quadrilateral faces increases by 9% (the vertex positions are the same as the original squares).

The smallest representative volume in a Kelvin foam contains seven faces, twelve

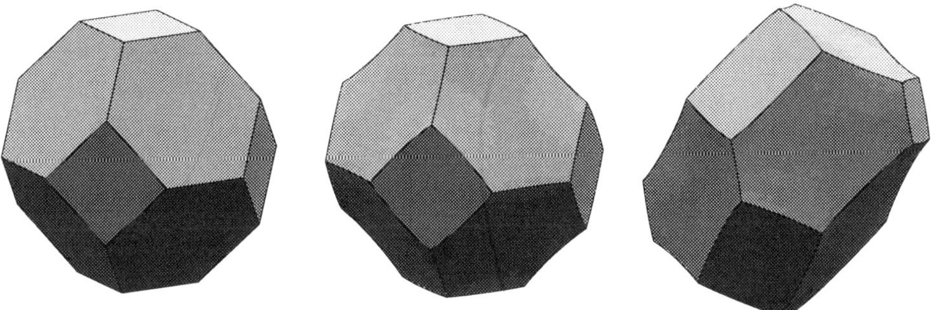

Fig. 1. The regular tetrakaidecahedron (truncated octahedron) has fourteen flat faces: six squares and eight hexagons. The Kelvin cell has flat quadrilateral faces and nonplanar hexagonal faces. The deformed Kelvin cell has seven faces, six edges, and three vertices that are unique. All faces are symmetric about their center.

edges, and six vertices. When the stress is isotropic, the cell centers are associated with a body-centered-cubic (bcc) lattice; all of the quadrilateral faces, hexagonal faces, edges, and vertices are identical to within a translation and rotation. By contrast, a deformed Kelvin cell has less symmetry; up to seven faces, six edges, and three vertices can be unique as each face is symmetric about its center (see Fig. 1). The lattice of the deformed foam differs from the original bcc lattice by a homogeneous displacement that corresponds to the macroscopic strain.

3. Volume Change

Pure volume deformation does not alter the shape of a Kelvin cell; the geometric expansion is self similar. Assuming ideal gas, the equation of state for a Kelvin foam is given by

$$P_f = \frac{P_b^0}{1+\Delta} - \frac{2E^0}{3(1+\Delta)^{1/3}} \tag{1}$$

which leads to the bulk modulus κ of linear elasticity

$$\kappa = P_b^0 - \frac{2}{9}E^0. \tag{2}$$

Here, P_f is the effective foam pressure, $(1 + \Delta)$ is the volume expansion ratio, Δ is the dilatation from classical elasticity theory, and $P_b{}^0$ and E^0 are the bubble pressure and interfacial energy density, respectively, of the undeformed foam ($\Delta = 0$). Under most circumstances (when the bubbles are sufficiently large) interfacial contributions can be neglected and both the foam pressure and bulk modulus are given by the bubble pressure.

The individual bubble pressures are different in disordered foams where cells have different topology; consequently, expansion cannot be self similar. Self-similar geometry is approached as $E/\overline{P}_b \to 0$, where $\overline{P}_b$ is the volume-average bubble pressure.

4. Extension

Consider uniaxial extension with no volume change where we pull on one quadrilateral face and push on the others, i.e. the stretching direction is perpendicular to a four-sided face. The evolution of cell geometry with Hencky strain ε for complete minimal surface solutions is shown in Fig. 2. The area of the pulled quadrilateral decreases with increasing strain but remains finite in the limit $\varepsilon \to \varepsilon_{T1} = 0.254$. At ε_{T1}, no stable solution satisfies Plateau's laws *and* maintains contact between original cell neighbors. This situation triggers a topological transition (T1), which marks the limit of reversible elastic behavior and the onset of plasticity. Topological transitions will be discussed later.

The entire stress tensor can be evaluated by averaging the local stress over a unit cell

$$\sigma = -P_b \boldsymbol{I} + \frac{2T}{V} \sum_{k=1}^{7} \int_{S_k} \left(\boldsymbol{I} - \boldsymbol{n}_k \boldsymbol{n}_k \right) dA \tag{3}$$

where $\boldsymbol{I}$ is the identity matrix, $\boldsymbol{n}_k$ is a unit vector normal to the k-th face, S_k refers to the k-th face, and dA is the differential area element. Alternatively, the tensile stress can be obtained from the surface energy using

$$\sigma_{zz} - \sigma_{xx} = \frac{dE}{d\varepsilon} = \frac{d}{d\varepsilon}\left(\frac{TS}{V} \right) \tag{4}$$

where S is the total interfacial area of a Kelvin cell. The stress-strain curve in Fig. 3 exhibits a maximum in the tensile stress and a turning point. The unstable solutions on the curve below the turning point have the same topology but higher surface area than their stable counterparts. These unstable solutions satisfy Plateau's laws, which are necessary but not sufficient criteria for stable structure (ALMGREN and TAYLOR, 1976; TAYLOR, 1976). The unstable solution at $\varepsilon = (1/3)\log 2$, where the tensile stress is zero, corresponds to a rhombic dodecahedron on a face-centered-cubic (fcc) lattice.

In view of the cost of computing minimal surfaces, it is enlightening to evaluate microrheological results based on oversimplified foam geometry. In all three approximations, the undeformed cell is a regular tetrakaidecahedron. The faces remain flat as the foam deforms but the dihedral angles are different in each case.

DERJAGUIN (1933) estimated the shear modulus of a random foam by assuming

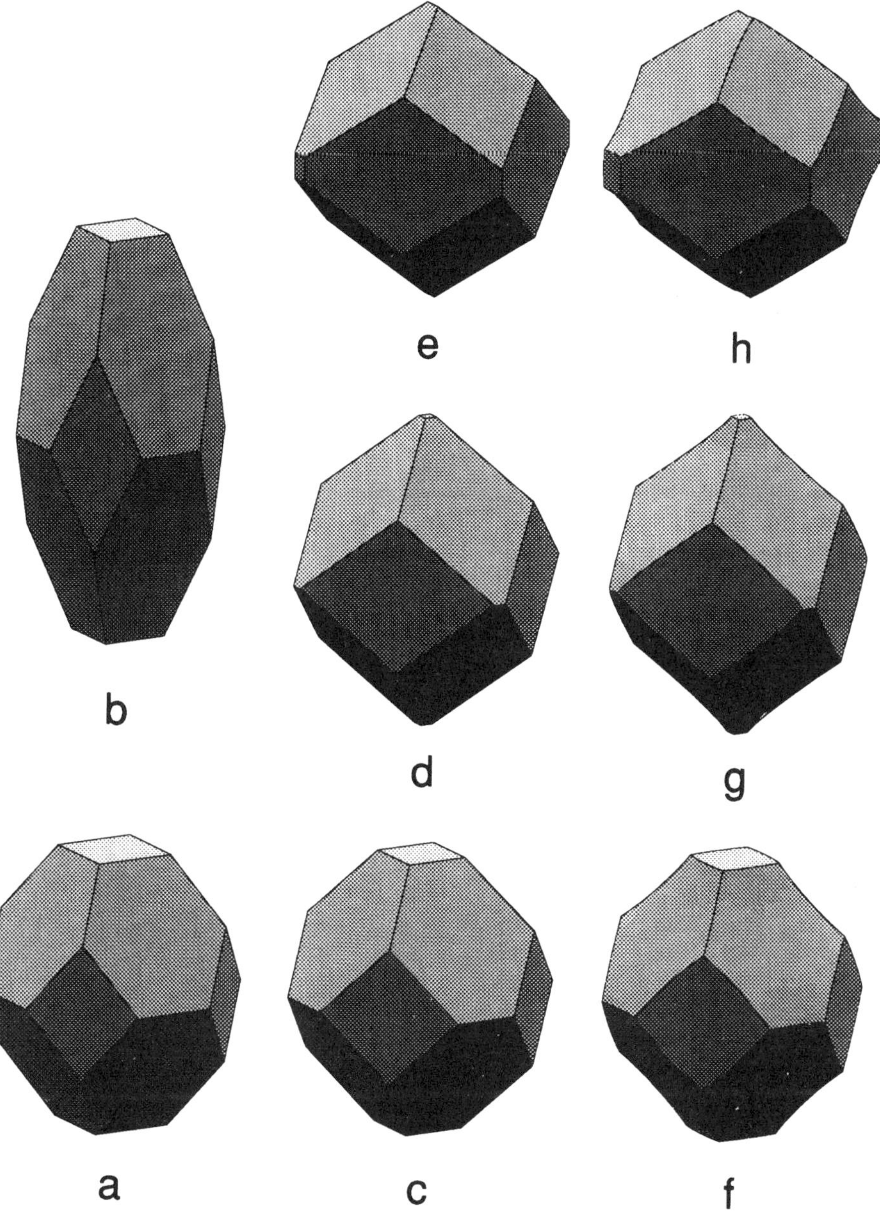

Fig. 2. The evolution of Kelvin cell geometry for uniaxial extension in a ⟨1,0,0⟩ direction. There is no topological transition for affine deformation of a regular tetrakaidecahedron: (a) $\varepsilon=0.1$, (b) 0.4. The minimal planar approximation for (c) 0.1, (d) 0.23, just before T1, (e) 0.26, after T1. Complete minimal surface results for (f) 0.1, (g) 0.254, just before T1, (d) 0.26, after T1. The cells lose symmetry after the point topological transitions.

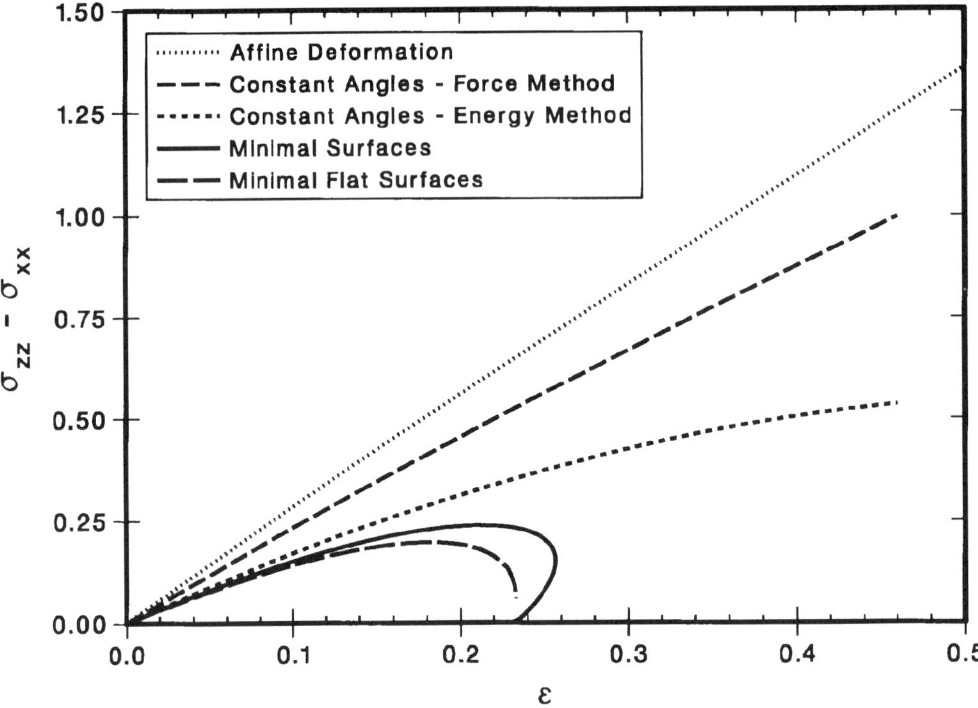

Fig. 3. Comparison of the stress-strain curves for the uniaxial extension in Fig. 2. The stress scale $T/V^{1/3}$ is the same in all figures.

affine deformation. With this assumption, our pulled square continues to shrink for all values of increasing strain (see Fig. 2). There is no criterion for topological transition so there is no elastic limit. The tensile stress goes as exp($\varepsilon/2$) for large ε. The affine assumption gives the largest tensile stress in Fig. 3.

Next, we assume constant dihedral angles that do not vary with strain. The pulled square shrinks to zero area when $\varepsilon = (2/3)\log 2$, causing eight edges to meet at some vertices. This violates the four-fold edge connectivity required by Plateau's laws and provides a T1 criterion. There are two different stress-strain curves in Fig. 3 that correspond to different methods for calculating the stress. The force and energy methods, which correspond to Eqs. (3) and (4), give inconsistent results in this situation where film-level forces are not balanced and energy is not minimized.

In the third approximation, dihedral angles are determined by minimizing the total surface area of flat faces (see Fig. 2). Like the complete solution, the pulled square shrinks but the strain at T1 is slightly smaller, $\varepsilon_{T1} = 0.233$. The tensile stress also exhibits a maximum and a turning point but notice that the curve for this minimal planar solution is below the complete solution (see Fig. 3). This indicates that approximate solutions for stress do not provide bounds for exact solutions.

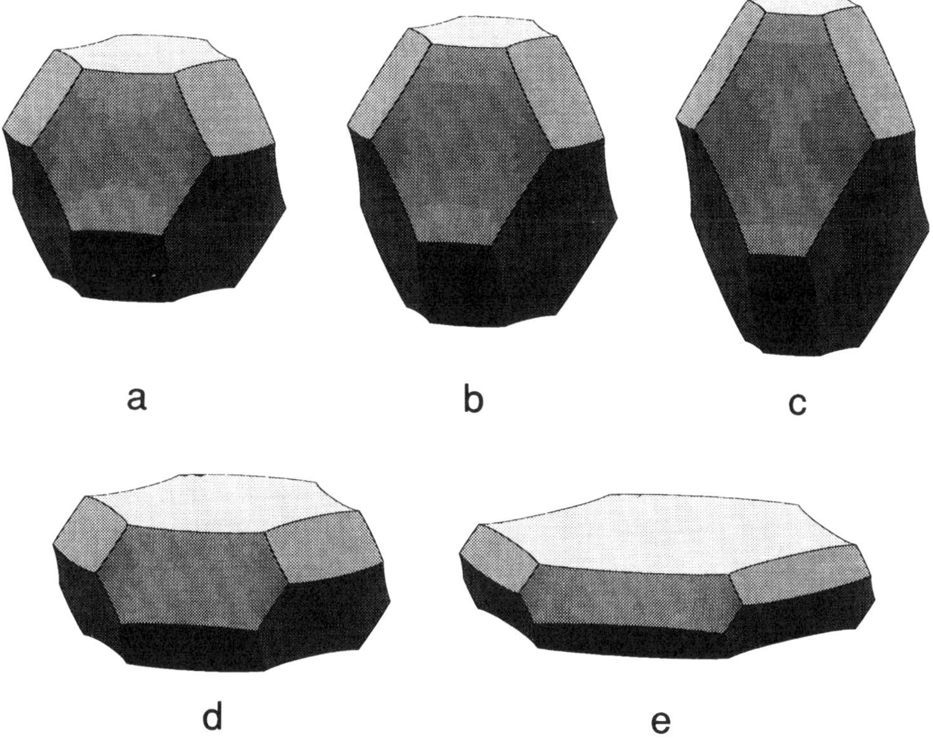

Fig. 4. The evolution of Kelvin cell geometry for uniaxial extension and compression in a ⟨1,1,1⟩ direction for which there are only two unique edges lengths and no topological transitions; (a) $\varepsilon = 0$, (b) 0.2, (c) 0.4, (d) –0.5, and (e) –1.

The Kelvin cell has cubic symmetry, which implies two shear moduli for linear elastic response. One of these, which happens to be the smaller, is calculated from the initial slope in Fig. 3 giving $G_{min} = 0.5706$ for the complete solution. The other shear modulus, $G_{max} = 0.9604$, can be evaluated by pulling on a hexagonal face; i.e. aligning the stretching direction with a ⟨1,1,1⟩ axis of the Kelvin cell (see Fig. 4). The disparity between G_{min} and G_{max} indicates significant anisotropic behavior. An effective isotropic shear modulus $\overline{G}$ is calculated by averaging the response over all possible foam orientations: $\overline{G} = (3/5)G_{max} + (2/5)G_{min} = 0.8070$. PRINCEN and KISS (1986) have measured the shear modulus of a series of concentrated oil-in-water emulsions with polydisperse drop-size distributions. For a dry monodisperse foam, their empirical result becomes $G = 0.821$, which is close to $\overline{G}$.

The elastic response is much less anisotropic for the elegant structure described by WEAIRE and PHELAN (1994), for which $G_{min} = 0.8538$, $G_{max} = 0.8902$, and $\overline{G} = 0.8684$. The Weaire-Phelan foam, with eight bubbles in the unit cell, is more random than a Kelvin

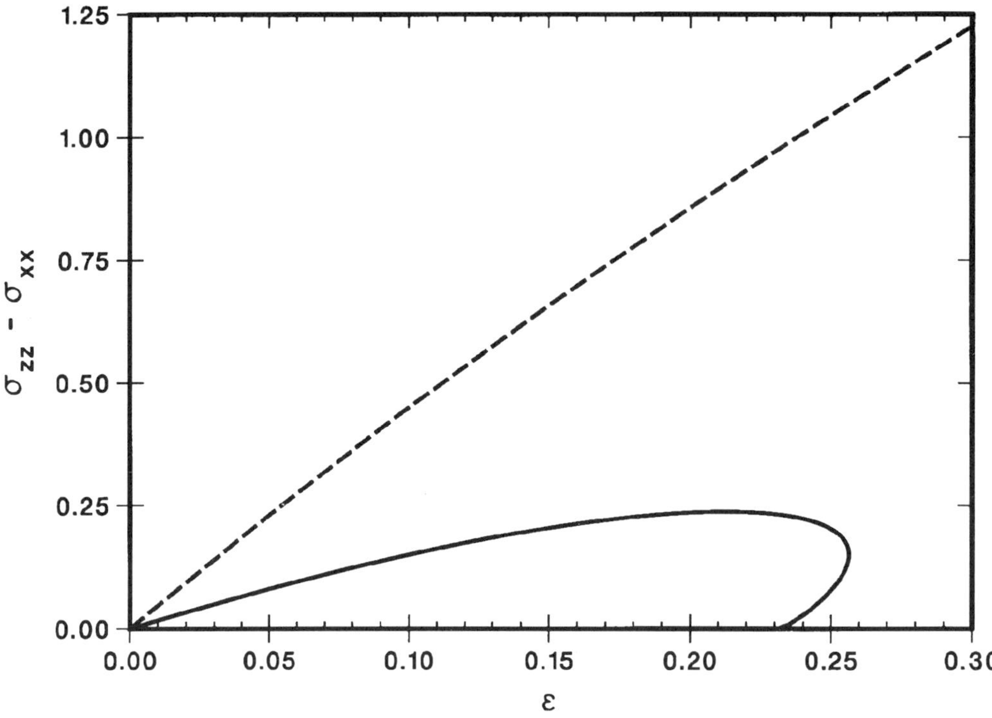

Fig. 5. Comparison of the stress-strain curves for uniaxial extension in a ⟨1,0,0⟩ direction (solid line) and ⟨1,1,1⟩ direction (dashed line). The corresponding cell geometries are shown in Figs. 2(f), 2(g) and 4.

cell but much less complex than other tetrahedrally close-packed (tcp) structures with over one hundred bubbles (RIVIER, 1994). We anticipate large tcp foams having nearly isotropic response with $\overline{G}$ approaching that of a random foam. The shear modulus of disordered foams has not been established on firm theoretical grounds. The *ad hoc* model of STAMENOVIC (1991), which considers local behavior in the vicinity of a foam vertex, leads to a very interesting formula, $G/E^0 = 1/6$, which gives $G = 0.8814$ for a Weaire-Phelan foam and $G = 0.8844$ for a Kelvin foam.

The stress-strain curves for the two different orientations of a Kelvin cell are qualitatively different for large deformations (see Fig. 5). There is no topological transition when pulling in a ⟨1,1,1⟩ direction (see Fig. 4). The absence of a T1 stems from high symmetry of the perfectly ordered structure, which only has two unique edge lengths in this situation. A disordered foam would exhibit isotropic elastic-plastic behavior with very frequent topological transitions.

5. Simple Shearing Flow

We now focus our attention on simple shearing flow. The Kelvin cell geometry and

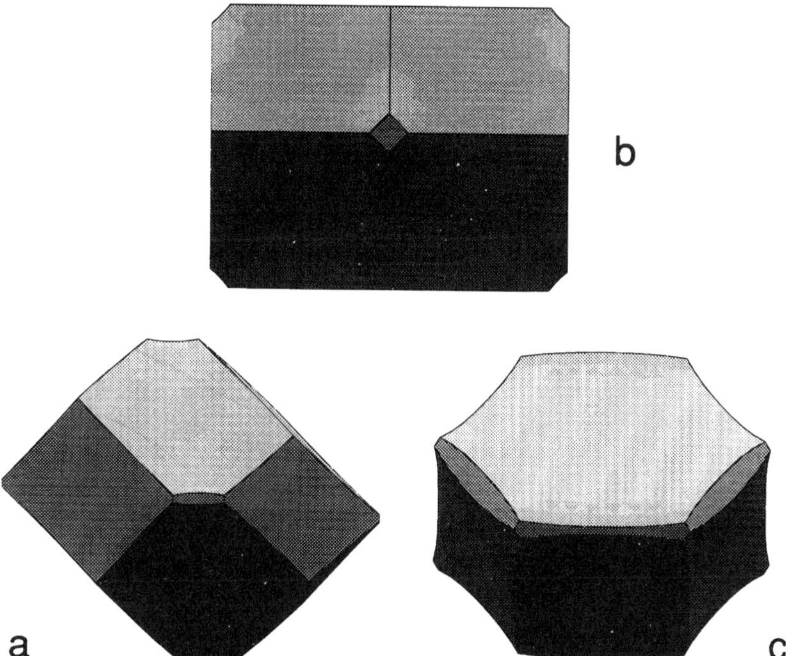

Fig. 6. Shrinking faces and edges on highly distorted Kelvin cells just prior to the three basic topological transitions: (a) standard T1, $\gamma = 0.60$, as in Figs. 7(c) and 8(c); (b) point T1, $\gamma = 0.54$, compare with Fig. 10(c); (c) triple T1, $\gamma = \sqrt{2}$, as in Fig. 10(g). The standard T1 for (a) occurs when the short edges go to zero length at $\gamma = 0.617$. The point T1 for (b) occurs at $\gamma = 0.548$. The triple T1 for (c) occurs when opposite "long" edges of the shrinking hexagonal face touch when $\gamma = 1.47$.

effective stress are piecewise continuous functions of the shear strain γ. Each discontinuity corresponds to a topological change that is preceded by shrinking faces. Each T1 reduces surface energy, results in cell-neighbor switching, and provides a cell-level mechanism for plastic yield behavior during foam flow. The foam structure can be determined for all strains by choosing initial foam orientations that lead to strain-periodic behavior.

Kelvin foam flow involves three basic T1 types: standard, point, and triple transitions (see Fig. 6). SCHWARZ (1965) observed and classified topological rearrangements in disordered polyhedral foams but only found one of these types, the standard T1. Conditions that trigger the point and triple transitions stem from the high symmetry of a Kelvin cell.

The cell orientation with the smallest strain period has two standard transitions per cycle. Figure 7 shows the evolution of cell geometry with strain while Fig. 8 gives a slightly different view of the first T1. Standard transitions occur when one of the six

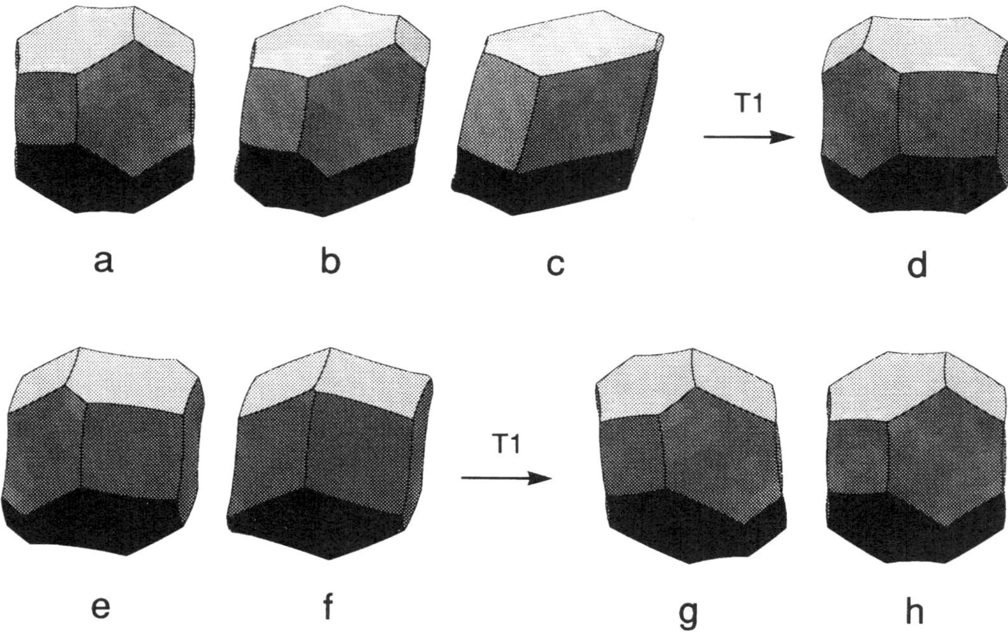

Fig. 7. Evolution of foam geometry with strain for the cell orientation that has the smallest strain period $\gamma_p = \sqrt{3/2}$: (a) $\gamma = 0$, (b) 0.30, (c) 0.60, (d) 0.62, (e) 0.80, (f) 0.99, (g) 1, (h) γ_p. Standard topological transitions occur when opposite edges of shrinking quadrilateral faces go to zero length at $\gamma = 0.617$ and $\gamma = 0.998$. The Kelvin cell in (a) and (h) has the same shape but some neighbors are different as a consequence of the T1s.

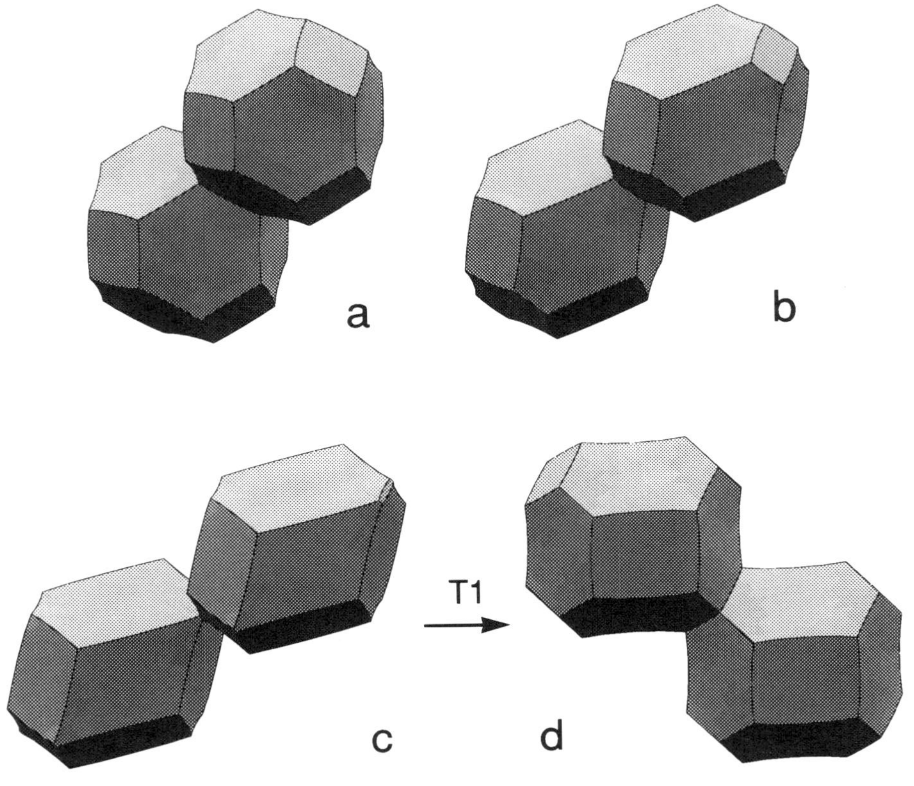

Fig. 8. Standard topological transition for the cell orientation in Fig. 7 with (a) $\gamma = 0$, (b) 0.30, (c) 0.60, and (d) 0.62. Cell neighbors in (a), (b), and (c) share the quadrilateral face that shrinks as γ increases. The T1 occurs when opposite edges of the quadrilateral face go to zero length at $\gamma = 0.617$. Original cell neighbors separate and a new face forms between the new neighbors in (d). Each cell is shown from a slightly different angle in Fig. 7. The view of (c) is also different in Fig. 6(a).

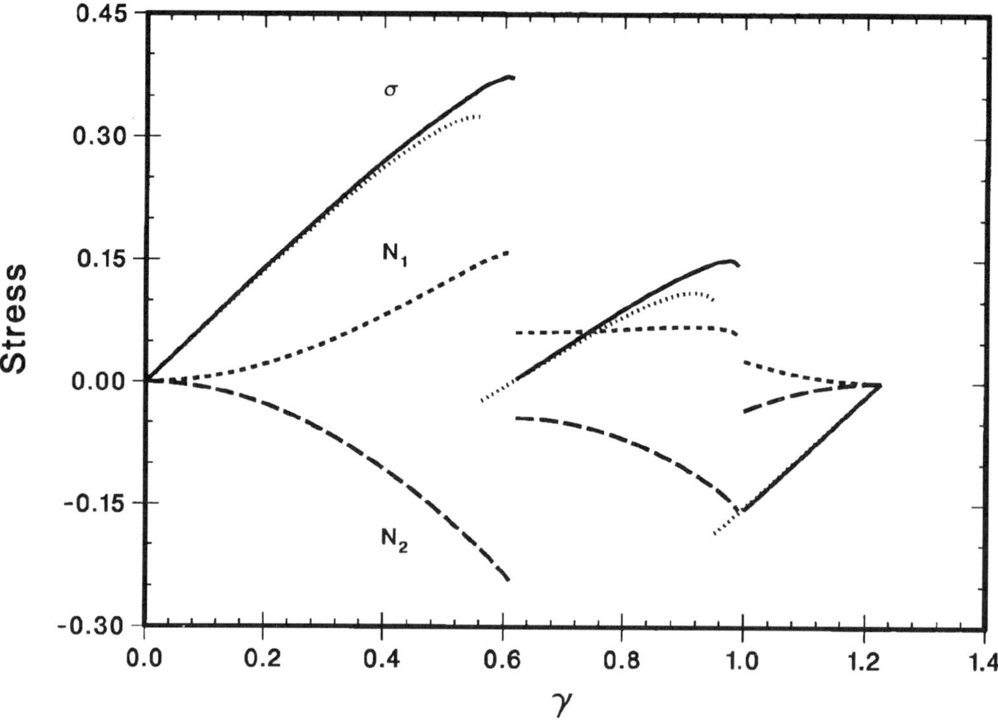

Fig. 9. Instantaneous stress as a function of shear strain γ for the situation in Figs. 7 and 8. The shear stress σ, first normal stress difference N_1, and second normal stress difference N_2 are periodic functions with strain period $\gamma_p = \sqrt{3/2}$. The dotted line corresponds to the shear stress for the minimal planar solution.

unique edge lengths tends to zero. The vanishing edge corresponds to opposite sides of a shrinking quadrilateral face (see Figs. 6–8). This face degenerates to form a single edge during the T1. The structure resulting from the T1 is still a Kelvin cell but all faces change shape; some quadrilaterals become hexagonal and *vice versa*. The cell neighbor that was attached to the vanishing face is lost but a neighbor is gained where a new quadrilateral face emerges from an edge. The instantaneous shear stress and normal stress differences are shown in Fig. 9 along with the shear stress for the minimal planar approximation.

A point transition occurs when two of the six unique edge lengths form all of the sides on a shrinking quadrilateral face and tend to zero simultaneously, as shown in Figs. 2, 6 and 10. Since there are two possible standard transitions associated with a quadrilateral face, depending on which set of opposite edges vanish, a point transition requires a choice between two structures that are mirror images.

A triple transition occurs when two unique shrinking edges are on different quadrilaterals (see Figs. 6 and 10). Two quadrilateral faces and a hexagonal face shrink simultaneously and all three corresponding cell neighbors are lost during the T1. Opposite

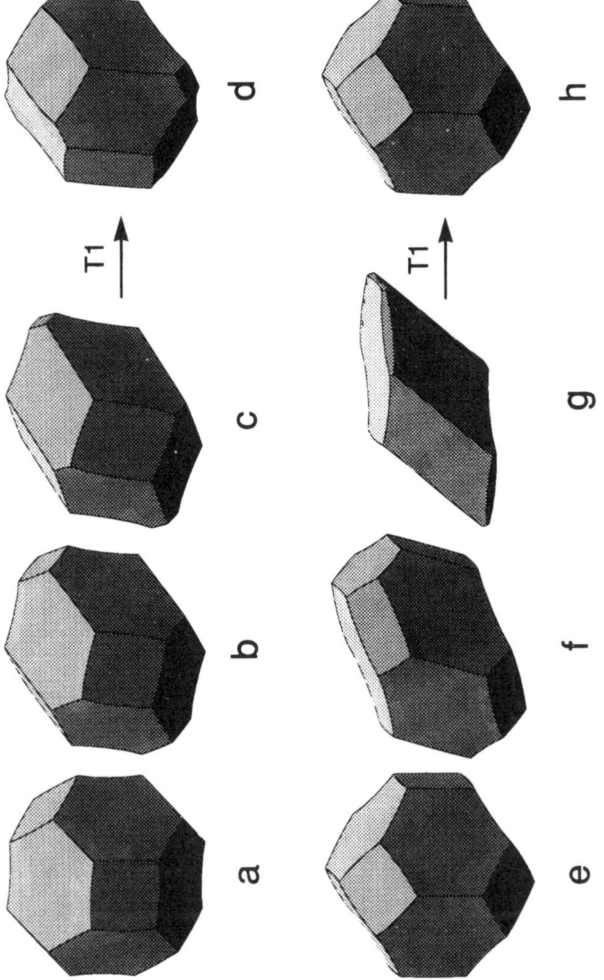

Fig. 10. Evolution of foam geometry with shear strain for an orientation with a point transition: (a) $\gamma = 0$, (b) 0.30, (c) 0.50, just before T1, and (d) 0.55, and an orientation with a triple transition (e) $\gamma = 0$, (f) 0.50, (g) $\sqrt{2}$, just before T1, and (h) 1.50. Topological transitions as in Figs. 6(b) and 6(c). Viewing angle as in Fig. 8.

edges of the shrinking hexagon that are not contracting bow in and eventually touch at their center triggering the triple T1.

Symmetry imposes strong restrictions on the outcome of T1s for a Kelvin foam. The foam topology changes through neighbor switching. The topology of individual faces changes as they gain or lose two sides. But Kelvin cells always beget Kelvin cells in a T1 process. By contrast, topological changes in foams without perfect order can alter the number of faces, the number of edges, and the distribution of n-sided faces on an individual cell and for the entire 3D foam. In 2D, the number of edges and vertices on a particular cell can change after a T1, but the total number of these features is fixed by Euler's law when there is three-fold edge connectivity.

Topological transitions in 2D foams are triggered by an edge length going to zero continuously with strain. The analogous 3D event—the area of a face tending to zero continuously with strain—does not occur in a Kelvin foam. Topological changes are initiated by one unique edge length going to zero, at a turning point where an edge length is small, or when opposite edges of a hexagonal face touch each other.

The stress-strain curves for simple shearing flow depend on foam orientation and exhibit large fluctuations as shown in Fig. 9. The dynamic (or viscometric) yield stress $\overline{\sigma}$ is calculated by averaging the shear stress σ over strain. This average stress varies by a factor of five for the three orientations represented in Figs. 7–10 ($\overline{\sigma}$ = 0.1191, 0.2040, 0.6168 when the first T1 is standard, point, triple).

6. Conclusions

The Kelvin cell enjoys special status in foam science as the fundamental entity in a perfectly ordered soap froth. But it is rarely found in real foams that have disordered structure (MATZKE, 1946) and exhibit isotropic rheological behavior. The cubic symmetry of a Kelvin foam guarantees isotropic elastic response to volume change but forecasts the anisotropic shear behavior that is observed.

The shear modulus of a Kelvin foam, averaged over all orientations, is close to measured values for real foams. But orientation dependence in the linear elastic regime is amplified at large strains. Consequently, the Kelvin cell is less useful for modeling the yield stress and other nonlinear phenomena associated with real foams.

The connection between rheology and geometric complexity can be established by starting with the Kelvin cell and progressing to tcp and disordered structures. Various research groups are now investigating the role of Plateau borders in wet foams and polydisperse cell-size distribution to gain systematic understanding of foam microrheology.

We thank Ken Brakke for developing and maintaining the Surface Evolver, which is available from the Geometry Center at the University of Minnesota (by anonymous FTP from geom.umn.edu). Interactions with Ken Brakke, Robert Phelan, Henry Princen, and Denis Weaire have guided our thoughts on this subject. DAR was supported by National Science Foundation Grant #CTS-9113907. AMK performed this work at Sandia National Laboratories with support from the U.S. Department of Energy under contract #DE-AC04-94AL85000.

Abstract. The microrheology of a Kelvin foam is investigated for large quasi-static deformations. The elastic-plastic response is determined by calculating the evolution of Kelvin cell geometry and effective stress with strain. The rheological response is highly anisotropic for uniaxial extension and simple shearing flow. The two shear moduli are $G_{min} = 0.5706$ and $G_{max} = 0.9646$, where stress is scaled by $T/V^{1/3}$, T is surface tension, and V is bubble volume. For large deformations, the foam structure and stress are piecewise continuous functions of strain. Each discontinuity corresponds to a topological transition (T1) that occurs when the minimal foam structure is unstable. Each T1 reduces surface energy, results in cell-neighbor switching, and provides a cell-level mechanism for plastic yielding. Computer simulations were performed with the Surface Evolver program developed by Brakke.

REFERENCES

ALMGREN, F. and TAYLOR, J. (1976) The geometry of soap films and soap bubbles, *Scientific American*, July, 82–93.

BOLTON, F. and WEAIRE, D. (1992) The effects of Plateau borders in the two-dimensional soap froth. II. General simulation and analysis of rigidity loss transition, *Philos. Mag. B*, **65**, 473–487.

BRAKKE, K. (1977) *The Motion of a Surface by Its Mean Curvature*, Princeton University Press.

BRAKKE, K. A. (1992) The surface evolver, *Experimental Mathematics*, **1**, 141–165.

BRAKKE, K. A. (1995) *Surface Evolver Manual*, version 1.99, July.

DERJAGUIN, B. V. (1933) Die elastischen eigenschaften der schaume, *Kolloid-Z.*, **64**, 1–6.

HERDTLE, T. (1991) Numerical studies of foam dynamics, Ph.D Thesis, University of California at San Diego, La Jolla, CA.

HUTZLER, S., WEAIRE, D. and BOLTON, F. (1995) The effects of Plateau borders in the two-dimensional soap froth III. Further results, *Philos. Mag. B*, **71**, 277–289.

KELVIN, Lord (Thompson, W.) (1887) On the division of space with minimum partitional area, *Philos. Mag.*, **24**, 503–514.

KHAN, S. A. and ARMSTRONG, R. C. (1986) Rheology of foams. I. Theory for dry foams, *J. Non-Newtonian Fluid Mech.*, **22**, 1–22.

KRAYNIK, A. M. (1988) Foam flows, *Ann. Rev. Fluid Mech.*, **20**, 325–357.

KRAYNIK, A. M. and HANSEN, M. G. (1986) Foam and emulsion rheology: A quasistatic model for large deformations of spatially periodic cells, *J. Rheology*, **30**, 409–439.

KRAYNIK, A. M. and REINELT, D. A. (1996a) Linear elastic behavior of dry soap foams, *J. Coll. Int. Sci.*, **181**, 511–520.

KRAYNIK, A. M. and REINELT, D. A. (1996b) The linear elastic behavior of a bidisperse Weaire-Phelan foam, *Chem. Eng. Comm.* (accepted).

KRAYNIK, A. M., REINELT, D. A. and PRINCEN, H. M. (1991) The nonlinear elastic behavior of polydisperse hexagonal foams and concentrated emulsions, *J. Rheology*, **35**, 1235–1253.

MATZKE, E. B. (1946) The three-dimensional shape of bubbles in foam—an analysis of the role of surface forces in three-dimensional cell shape determination, *Am. J. Botany*, **33**, 58–80.

PLATEAU, J. A. F. (1873) *Statique Experimentale et Theorique des Liquides Soumis aux Seules Forces Moleculaires*, Gauthier-Villiard.

PRINCEN, H. M. (1983) Rheology of foams and highly concentrated emulsions. I. Elastic properties and yield stress of a cylindrical model system, *J. Coll. Int. Sci.*, **91**, 160–175.

PRINCEN, H. M. and KISS, A. D. (1986) Rheology of foams and highly concentrated emulsions. III. Static shear modulus, *J. Coll. Int. Sci.*, **112**, 427–437.

REINELT, D. A. (1993) Simple shearing flow of three-dimensional foams and highly concentrated emulsions with planar films, *J. Rheology*, **37**, 1117–1139.

REINELT, D. A. and KRAYNIK, A. M. (1989) Viscous effects in the rheology of foams and concentrated emulsions, *J. Coll. Int. Sci.*, **132**, 491–503.

REINELT, D. A. and KRAYNIK, A. M. (1990) On the shearing flow of foams and concentrated emulsions, *J. Fluid Mech.*, **215**, 431–455.

REINELT, D. A. and KRAYNIK, A. M. (1993) Large elastic deformations of three-dimensional foams and highly concentrated emulsions, *J. Coll. Int. Sci.*, **159**, 460–470.

REINELT, D. A. and KRAYNIK, A. M. (1996) Simple shearing flow of a dry Kelvin foam, *J. Fluid Mech.*, **311**, 327–343.

RIVIER, N. (1994) Kelvin's conjecture on minimal froths and the counter-example of Weaire and Phelan, *Phil. Mag. Lett.*, **69**, 297–303.

SCHWARTZ, L. W. and PRINCEN, H. M. (1987) A theory of extensional viscosity for flowing foams and concentrated emulsions, *J. Coll. Int. Sci.*, **118**, 201–211.

SCHWARZ, H. W. (1965) Rearrangements in polyhedric foam, *Recueil*, **84**, 771–781.

STAMENOVIC, D. (1991) A model of foam elasticity based upon the laws of Plateau, *J. Coll. Int. Sci.*, **145**, 255–259.

TAYLOR, J. E. (1976) The structure of singularities in soap-bubble-like and soap-film-like minimal surfaces, *Ann. Math.*, **103**, 489–539.

WEAIRE, D. and FORTES, M. A. (1994) Stress and strain in liquid and solid foams, *Advances in Physics*, **43**, 685–738.

WEAIRE, D. and FU, T.-L. (1988) The mechanical behavior of foams and emulsions, *J. Rheology*, **32**, 271–283.

WEAIRE, D. and KERMODE, J. P. (1983) Computer simulation of a two-dimensional soap froth, *Philos. Mag. B*, **48**, 245–259.

WEAIRE, D. and KERMODE, J. P. (1984) Computer simulation of a two-dimensional soap froth. II. Analysis of results, *Philos. Mag. B*, **50**, 379–395.

WEAIRE, D. and PHELAN, R. (1994) A counter-example to Kelvin's conjecture on minimal surfaces, *Phil. Mag. Lett.*, **69**, 107–110.

Close-Packing and Froth*

H. S. M. COXETER

University of Toronto, Toronto, Canada

Cannon-balls may aid the truth,
But thought's a weapon stronger;
We'll win our battles by its aid;—
Wait a little longer.
CHARLES MACKAY (1814–1889)
("The Good Time Coming")

1. Algebraic Introduction

The abstract groups $(2, p, q)$, defined by

$$R^p = S^q = (RS)^2 = 1,$$

or

$$R^p = S^q = T^2 = RST = 1,$$

or

$$S^q = T^2 = (ST)^p = 1,$$

have been studied intensively ever since HAMILTON (1856) expressed $(2, 3, 5)$ in the form

$$\iota^2 = \kappa^3 = \lambda^5 = 1, \qquad \lambda = \iota\kappa$$

*Reproduced from *Illinois Journal of Mathematics* (1958) Vol. **2**, No. 4B, Miller Memorial Issue.
Keywords: Froth, Close-Packing, Characteristic Triangle, Characteristic Tetrahedron (given by the editors of FORMA).

and wrote, "I am disposed to give the name 'Icosian Calculus' to this system of symbols." DYCK (1882, p. 35; see also BURNSIDE, 1911, p. 407) expressed the symmetric and alternating groups

$$\mathcal{S}_3,\ \mathcal{A}_4,\ \mathcal{S}_4,\ \mathcal{A}_5$$

in the form (2, 3, q) with $q = 2, 3, 4, 5$, respectively. MILLER (1935–59a, p. 117) remarked that the case when $q = 6$ is entirely different. In fact (MILLER, 1935–59b), the group (2, p, q) is finite if and only if

$$(p-2)(q-2) < 4. \tag{1}$$

Thus the finite groups in the family are

the dihedral group (2, 2, q) of order $2q$,
the tetrahedral group (2, 3, 3) of order 12,
the octahedral group (2, 3, 4) of order 24,
the icosahedral group (2, 3, 5) of order 60.

The inequality (1) is a necessary and sufficient condition for the finiteness of the number c, which we define to be the period of any one of the elements

$$R^2S^2,\ R^{-1}S^{-1}RS,\ TSR,\ S^{-1}TST.$$

In the infinite case, BRAHANA (1928) obtained interesting factor groups by assigning a finite period to these elements.

When (2, 3, q) is expressed in the form

$$S^q = T^2 = (ST)^3 = 1,$$

the commutator $TS^{-1}TS = STS{\cdot}S$ is conjugate to S^3T; thus c is the period of S^3T. If $q = 3$, we have $S^3T = T$, so that $c = 2$. If $q = 4$, $S^3T = (TS)^{-1}$, so that $c = 3$. If $q = 5$, S^3T is conjugate to $S^{-1}TS^{-1} = TST$, so that $c = 5$. In Section 5 we shall find a natural way to combine these three results in the single formula

$$c = \frac{12}{7-q} - 1 \quad (2 < q < 6)$$

and to express the order of (2, 3, q) in the form $2c(c+1)$.

In Section 6 we shall obtain the criterion

$$p - 4/p + 2q + r - 4/r < 12$$

for the finiteness of the group (TODD, 1931, p. 217)

$$R^p = S^q = T^r = (RS)^2 = (RST)^2 = (ST)^2 = 1 \quad (p, q, r > 2). \tag{2}$$

2. Geometric Introduction

Hamilton's "Icosian Calculus" and Klein's "Lectures on the Icosahedron" remind us that one of the most significant properties of the polyhedral groups (2, p, q) is their occurrence as finite groups of rotations in three-dimensional Euclidean (or non-Euclidean) space, i.e., as the rotation groups of the Platonic solids $\{p, q\}$. In the words of the late WEYL (1952, p. 78, 79), "These groups ... are an immensely attractive subject for geometric investigation."

In the Schläfli symbol $\{p, q\}$, p is the number of vertices (or sides) of a face, and q is the number of faces (or edges) at a vertex; e.g., the cube is $\{4, 3\}$. If such a polyhedron has V vertices, E edges, and F faces, we easily verify that

$$qV = 2E = pF$$

(COXETER, 1948, p. 11). With the aid of Euler's formula $V - E + F = 2$, we can deduce expressions for V, E, F as functions of p and q. For instance, if $q = 3$,

$$V = \frac{4p}{6-p}, \qquad E = \frac{6p}{6-p}, \qquad F = \frac{12}{6-p}. \tag{3}$$

In (4) we shall see why E is always the product of two consecutive integers.

A *Petrie polygon* of $\{p, q\}$ is a skew $2c$-gon whose sides are $2c$ edges of the solid, so chosen that any two consecutive sides, but no three, belong to a face; e.g., the Petrie polygon of a cube is a skew hexagon. Every edge belongs to two faces, and therefore also to two Petrie polygons. Since there are E edges, there are E/c Petrie polygons.

Since a Petrie polygon is symmetrical by a half-turn about the line joining the midpoints of two opposite sides, the midpoints of its $2c$ sides lie in a plane and are the vertices of a plane $2c$-gon, $\{2c\}$.

Reciprocating $\{p, q\}$ with respect to the sphere that touches its edges, we obtain the *reciprocal* polyhedron $\{q, p\}$, which has F vertices, E edges, and V faces. Its edges cross those of $\{p, q\}$ at right angles. Thus the Petrie polygon of $\{q, p\}$ is of the same type as that of $\{p, q\}$, and the plane polygon formed by the midpoints of its sides is the same $\{2c\}$.

In Section 5 we shall follow the procedure of STEINBERG (1958) to obtain an explicit formula for c.

A *vertex figure* of $\{p, q\}$ is the plane q-gon, $\{q\}$, whose vertices are the midpoints of the q edges meeting at one vertex, i.e., the section of the solid by the plane through these midpoints. Thus $\{p, q\}$ may be described as having face $\{p\}$ and vertex figure $\{q\}$.

The four-dimensional analogues of the five Platonic solids are the six regular hypersolids, which include the regular simplex $\{3, 3, 3\}$ and the hypercube $\{4, 3, 3\}$. Such a four-dimensional polytope $\{p, q, r\}$ is a configuration of equal polyhedra $\{p, q\}$, called cells, fitting together in such a way that each face $\{p\}$ belongs to two cells, and each edge to r cells. It follows that the arrangement of the cells at a vertex corresponds to the arrangement of the faces of a $\{q, r\}$, in the sense that each face of the $\{q, r\}$ is a vertex figure of the corresponding cell. This $\{q, r\}$, whose vertices are the midpoints of the edges at one vertex of $\{p, q, r\}$, is naturally called the *vertex figure* of the polytope (COXETER, 1948, p. 129). Thus $\{p, q, r\}$ may be described as having cell $\{p, q\}$ and vertex figure $\{q, r\}$.

In Section 6 we shall obtain a new version (7) of one of the principal results in *Regular Polytopes* (COXETER, 1948, p. 232), and use it to obtain a criterion for the available values of p, q, r.

The notation $\{p, q, r\}$ extends naturally from finite polytopes to infinite honeycombs so as to suggest the symbol $\{4, 3, 4\}$ for the three-dimensional honeycomb of cubes, whose vertices may be taken to be all the points (x, y, z) for which x, y, z are integers. Its cell is the cube $\{4, 3\}$, and its vertex figure is the octahedron $\{3, 4\}$ whose 8 faces are the vertex figures of the 8 cubes that surround a vertex.

In Section 7 we shall attempt to describe a hypothetical regular honeycomb $\{p, 3, 3\}$, which has a value of p lying between 5 and 6 but still may be regarded as existing in a statistical sense. It provides a theoretical explanation for some experimental results obtained by Matzke and his colleagues.

3. Close-Packing

In old war memorials one often sees a pyramidal heap of cannon balls: one at the top resting on four others which, in turn, rest on nine, and so on, the n-th horizontal layer containing n^2. Each interior ball touches twelve others: four in its own layer, four above, and four below. The centers of these twelve spheres are the vertices of a cuboctahedron, i.e., they are the mid-edge points of a cube. The shape of the whole square pyramid is just the "top" half of a regular octahedron, since each sloping face is an equilateral triangle formed by $1 + 2 + 3 + \cdots$ cannon balls.

This triangular arrangement suggests the simpler problem of packing equal circles in a plane (FEJES TÓTH, 1953, p. 58), or stacking circular cylinders. Each circle touches six others whose centers are the vertices of a regular hexagon. In other words, the circles are the incircles of the cells of the regular tessellation $\{6, 3\}$, which has three hexagons at each vertex.

One way to pack equal spheres is to begin with a horizontal layer whose "equators" form such a packing of circles. The same arrangement in the next layer above can be so placed that each sphere rests on three, and this can be done in two equivalent ways. When we come to the third layer, the two ways are no longer equivalent (STEINHAUS, 1950,

p. 170; HILBERT and COHN-VOSSEN, 1952, p. 46): in *hexagonal* close-packing each sphere in the third layer is exactly above one in the first layer, but in *cubic* close-packing the repetition is delayed till the fourth layer. It was BARLOW (1883) who first pointed out that the latter is the same as the normal piling of cannon balls (after the application of a suitable rotation). Hexagonal close-packing and cubic close-packing are equally dense, but the latter is more nearly "isotropic" since the spheres occur in straight rows in six different directions: the centers of the spheres form a *lattice* (in the crystallographic sense). Cubic close-packing is actually the densest lattice-packing. Gauss's original proof has been simplified by MORDELL (1948) and DEMPSTER (1957).

This lattice is called the "face-centered cubic lattice", because it can be derived from the simple cubic lattice {4, 3, 4} by taking not only the vertices but also the centers of the square faces. In other words, the spheres are the inspheres of the cells of the honeycomb of rhombic dodecahedra (STEINHAUS, 1950, p. 153).

The densest lattice-packing is not necessarily the densest packing. A first suspicion in this direction arises from the existence of equally dense non-lattice packings: the hexagonal close-packing and also several hybrids (MELMORE, 1947). This suspicion is increased by carefully examining the twelve neighbors of any one sphere. In the lattice-packing their centers are the vertices of a cuboctahedron (the reciprocal of the rhombic dodecahedron), whose faces consist of 8 triangles and 6 squares. Another familiar polyhedron having 12 vertices is the regular icosahedron, whose faces are 20 triangles. If a sphere is surrounded by 12 equal spheres located in this manner, the 12 spheres, while all touching the first, will not touch one another at all. Accordingly, if we let them roll on the first sphere until they are concentrated in one direction, it seems plausible that they might somehow make room for one more, so that the first sphere would touch thirteen others. According to H. W. Turnbull, who has studied the unpublished notebooks of David Gregory, this idea originated in a conversation of Gregory with Newton about 1694, apropos of the distribution of stars of various magnitudes! It remained an open question till 1953, when its impossibility was established by SCHÜTTE and VAN DER WAERDEN (1953); see also LEECH (1956). Although the thirteenth sphere cannot quite touch the one in the middle, it can be pushed in far enough to make a hopeful beginning for a dense packing that might conceivably be continued. BOERDIJK (1952) describes such a packing in an infinite tubular region, but there is apparently no satisfactory way to fill space by stacking such regions.

Stephen Hales stated, in his *Vegetable Staticks* (HALES, 1727, p. 95, 206): "I compressed several fresh parcels of Pease in the same Pot, with a force equal to 1600, 800, and 400 pounds; in which Experiments, tho' the Pease dilated, yet they did not raise the lever, because what they increased in bulk was, by the great incumbent weight, pressed into the interstices of the Pease, which they adequately filled up, being thereby formed into pretty regular Dodecahedrons."

MARVIN (1939) and MATZKE (1950) repeated Hales's experiment, replacing his peas by lead shot, "carefully selected under a microscope for uniformity of size and shape", in a steel cylinder, compressed with a steel plunger at a sufficient pressure (about 40,000 pounds) to eliminate all interstices. They found that, if the shot were stacked in cannonball fashion and compressed, nearly perfect rhombic dodecahedra were formed. But "if the shot were just poured into the cylinder the way Hales presumably put his peas into the iron pot, irregular 14-faced bodies were formed .... They were never rhombic dodecahedra." Nearly all the faces were either quadrangles, pentagons, or hexagons, with pentagons predominating.

HULBARY (1948) examined cells in undifferentiated vegetable tissues, and concluded that the internal cells have an average of approximately 14 faces. Among 650 such cells, chosen without special selection, he found a remarkable variety of shapes. The most prevalent (32 of the 650) had only 13 faces: 3 quadrangles, 6 pentagons, 4 hexagons; 33 edges, and 22 vertices.

4. Froth

LORD KELVIN (1887) believed that, of the various polyhedra which can be repeated to fill Euclidean space without interstices, the shape with the smallest surface for its volume is the truncated octahedron, whose faces consist of 8 hexagons and 6 squares. (For this solid he coined the outrageous name "tetrakaidecahedron", as if it were the chief or only polyhedron having fourteen faces! Actually it is one of the thirteen Archimedean solids, and it appears as one of the perspective drawings of models made by Leonardo da Vinci in Fra Luca Paccioli's *Divina proportione* (PACIOLI, 1509, 1946, p. 240). The name *truncum octaëdron* is due to Kepler.) Kelvin's conjecture is supported by the fact that, if S is the surface and C the volume, the value of S^3/C^2 is 150.1 ... for the truncated octahedron, and 152.8 ... for the rhombic dodecahedron (FEJES TÓTH, 1953, p. 174).

In the space-filling of truncated octahedra (whose centers form the "body-centered cubic lattice", consisting of the vertices and cell-centers of the simple cubic lattice), there are three cells at each edge, four at each vertex. To this extent it agrees with the theoretical specification for a froth of approximately uniform bubbles. But the balancing of surface tensions requires equal angles of 120° between the three faces that come together at an edge. Seeing that all the dihedral angles of the truncated octahedron are different from 120° (some greater, some less), Kelvin proposed a modification in which the flat hexagons are replaced by monkey-saddle-shaped minimal surfaces (cf. HILBERT and COHN-VOSSEN, 1952, p. 192), where however, Fig. 200 is obviously incorrect).

This "remarkable conformation" (THOMPSON, 1952, p. 552) remained unchallenged till 1940, when Matzke made a microscopic examination of an actual froth of 1900 measured bubbles, each one-tenth of a cc. (MATZKE, 1950, p. 225). "For 600 central bubbles examined, the average number of contacts was 13.70 ..., however, not a single bubble ... had the configuration which Kelvin had described and which Thompson had accepted. The commonest combination was 1-10-2 (118 of 600 bubbles). There were no rhombic dodecahedra". ("1-10-2" means "1 quadrangle, 10 pentagons, 2 hexagons".)

5. The Platonic Solids and Their Characteristic Triangles

The planes of symmetry of the regular polyhedron $\{p, q\}$ meet a concentric sphere

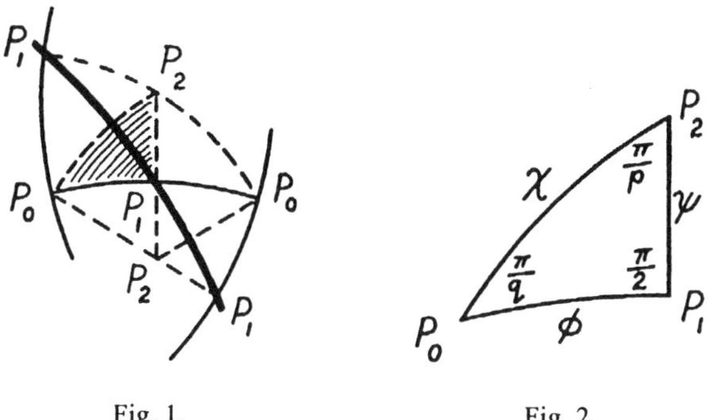

Fig. 1. Fig. 2.

in great circles which we shall call *circles of symmetry*. They decompose the sphere into $4E$ congruent right-angled spherical triangles $P_0P_1P_2$, where P_0 is on the radius through a vertex, P_1 is on the radius through the midpoint of an edge, and P_2 is on the radius through the center of a face (COXETER, 1948, p. 24).

We are assuming here that p and q are integers, greater than 2, satisfying (1), so that the polyhedron $\{p, q\}$ is nondegenerate. Clearly, the reciprocal polyhedron $\{q, p\}$ yields the same network of spherical triangles with the symbols P_0 and P_2 interchanged.

Figure 1 shows six such *characteristic triangles* in the neighborhood of an edge P_0P_0. The three points marked P_1, being the midpoints of three consecutive sides of a Petrie polygon, are three consecutive vertices of a $\{2c\}$ (see Section 2). The great circle containing these $2c$ points P_1 is called an *equator* ((COXETER, 1948, p. 67), where $2c$ is denoted by h). There are E/c equators: one for each Petrie polygon. Since the

$$\frac{E}{c}\left(\frac{E}{c}-1\right)$$

points of intersection of pairs of equators are just the E points P_1 (which occur in $(1/2)E$ pairs of antipodes), we have

$$\frac{E}{c}\left(\frac{E}{c}-1\right)=E,$$

whence

$$E=c(c+1). \tag{4}$$

Any equator is crossed by each circle of symmetry in a pair of antipodal points. It is crossed twice at each point P_1 (by two perpendicular circles of symmetry) and once at the midpoint of each arc P_1P_1 (by the hypotenuse P_0P_2 of a triangle $P_0P_1P_2$). Hence the number

Fig. 3.

of circles of symmetry (or of planes of symmetry) is $3c$; cf. COXETER (1948, p. 68).

Figure 3 shows the 9 circles of symmetry (light) and the 4 equators (dark) for the cube $\{4, 3\}$ or the octahedron $\{3, 4\}$, drawn in stereographic projection; cf. BURNSIDE (1911, frontispiece).

The $3c(3c - 1)$ points of intersection of pairs of circles of symmetry consist of $(1/2)p(p - 1)$ at each of the F points P_2, $(1/2)q(q - 1)$ at each of the V points P_0, and one at each of the E points P_1. Hence

$$\begin{aligned} 3c(3c-1) &= \frac{1}{2}p(p-1)F + \frac{1}{2}q(q-1)V + E \\ &= (p-1)E + (q-1)E + E \\ &= (p+q-1)c(c+1), \end{aligned}$$

and therefore

$$c + 1 = \frac{12}{10 - p - q}. \tag{5}$$

The connection with the c of Section 1 is seen by writing

$$R = R_1 R_2, \qquad S = R_2 R_3, \qquad T = R_3 R_1,$$

where R_1, R_2, R_3 are the reflections in the sides P_1P_2, P_2P_0, P_0P_1 of the characteristic triangle. Since $2c$ is the period of the product $R_1R_2R_3$ (COXETER, 1948, p. 91), c itself is the period of

$$(R_1 R_2 R_3)^2 = RTS = R^2 S^2.$$

The spherical triangle $P_0P_1P_2$ has angles π/p at P_2, $\pi/2$ at P_1, π/q at P_0. Let the respectively opposite sides be denoted by ϕ, χ, ψ, as in Fig. 2 (COXETER, 1948, p. 24). Since the sides of all the $4E$ triangles are arcs of the $3c$ circles of symmetry, each described twice, we have

$$4E(\phi + \chi + \psi) = 6c \cdot 2\pi,$$

whence

$$\phi + \chi + \psi = \frac{3c\pi}{E} = \frac{3\pi}{c+1} = \frac{(10 - p - q)\pi}{4} \qquad (6)$$

(COXETER, 1948, p. 74).

6. The Regular Hypersolids and Their Characteristic Tetrahedra

The hyperplanes of symmetry of a finite polytope $\{p, q, r\}$ (that is, the hyperplanes which act as "mirrors" reflecting the polytope into itself) meet a concentric hypersphere in "great spheres" which we naturally call *spheres of symmetry*. They decompose the hypersphere into (say) g congruent quadrirectangular spherical tetrahedra $P_0P_1P_2P_3$, whose four vertices are on the radii through a vertex, the midpoint of an edge, the center of a face, and the center of a cell (COXETER, 1948, p. 139). This *characteristic tetrahedron* is said to be "quadrirectangular" because all four faces are right-angled triangles. It can most easily be visualized by comparing it with the Euclidean tetrahedron that arises from the analogous simplicial subdivision of the cubic honeycomb $\{4, 3, 4\}$ in ordinary space (COXETER, 1948, p. 71), where the edges P_0P_1, P_1P_2, P_2P_3 are not only mutually orthogonal (as they always must be) but straight, and equal in length (since in this case P_0P_1 is half an edge of a cube, P_1P_2 is the inradius of a square face, and P_2P_3 is the inradius of the whole cube). Figure 4 shows an unfolded "net" which the reader may like to copy on thick paper, cut out, and fold up to make a solid model.

The angles between the edges P_3P_0, P_3P_1, P_3P_2 are equal to the sides of the char-

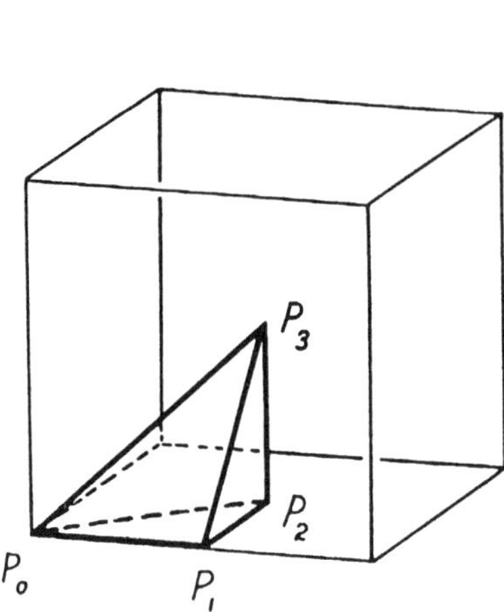

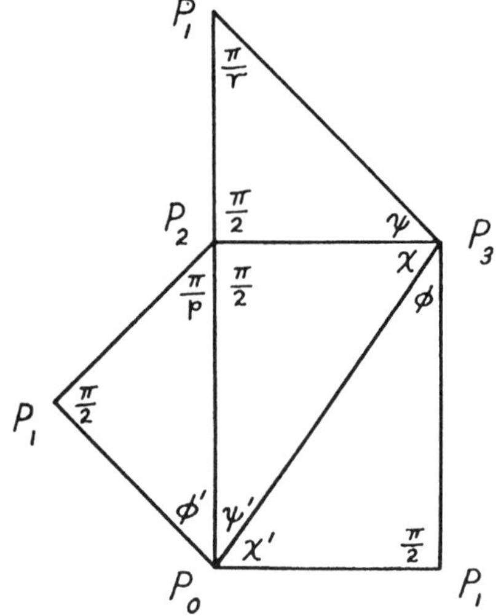

Fig. 4.

acteristic triangle for the cell $\{p, q\}$, namely

$$\angle P_0P_3P_1 = \phi, \quad \angle P_0P_3P_2 = \chi, \quad \angle P_1P_3P_2 = \psi.$$

Since P_2 is the center of a spherical p-gon of which P_0P_1 is half a side, $\angle P_0P_2P_1 = \pi/p$. Since the face $P_0P_1P_2$ is perpendicular to the edge P_3P_2, $\angle P_0P_2P_1$ can alternatively be obtained as the dihedral angle at this edge, which is the angle π/p of the characteristic triangle for $\{p, q\}$.

Similarly, the angles between the edges P_0P_1, P_0P_2, P_0P_3 are equal to the sides (say ϕ', χ', ψ') of the characteristic triangle for the vertex figure $\{q, r\}$, namely

$$\angle P_1P_0P_2 = \phi', \quad \angle P_1P_0P_3 = \chi', \quad \angle P_2P_0P_3 = \psi'.$$

Since the face $P_1P_2P_3$ is perpendicular to the edge P_0P_1, $\angle P_2P_1P_3$ is equal to the dihedral angle at this edge, which is the angle π/r of the characteristic triangle for $\{q, r\}$.

Each sphere of symmetry is tessellated with triangular faces of characteristic tetrahedra. Let a denote the average number of such triangles covering a sphere of symmetry. (This is actually the precise number of triangles, since every sphere of symmetry contains the same number; see STEINBERG (1959, Corollary 5.2), where a is denoted by g/h, as in (COXETER, 1948, p. 231).) Since each of the g tetrahedra has 4 faces, and each face belongs to 2 tetrahedra, the total number of triangles is $2g$, and the number of spheres of symmetry is $2g/a$.

In terms of the radius of the hypersphere as unit, the total area of the $2g/a$ great spheres is $8\pi g/a$. This must be equal to the sum of the angular excesses of the $2g$ spherical triangles. Since the sum of all the angles of all the triangles is $(1/2)g$ times the sum of the twelve face-angles of a single tetrahedron (Fig. 4), we can use (6) to obtain

$$\begin{aligned} 8\pi g / a &= \frac{1}{2} g(\phi + \chi + \psi + \phi' + \chi' + \psi' + \pi / p + \pi / r + 4\pi / 2) - 2g\pi \\ &= \frac{1}{2} g \cdot \frac{1}{4} \pi(10 - p - q + 10 - q - r + 4 / p + 4 / r + 8 - 16) \\ &= \frac{1}{8} \pi g(12 - p - 2q - r + 4 / p + 4 / r), \end{aligned}$$

whence (STEINBERG, 1958)

$$64 / a = 12 - (p - 4 / p) - 2q - (r - 4 / r). \tag{7}$$

Since a is positive for a four-dimensional polytope and infinite for a three-dimensional honeycomb, we must have

$$p - 4 / p + 2q + r - 4 / r \leq 12,$$

with equality only in the case of a honeycomb, namely when

$$p = r = 4 \quad \text{and} \quad q = 3.$$

It is interesting to observe how this algebraic criterion has the same effect as Schläfli's trigonometrical criterion (SCHLÄFLI, 1950, p. 215)

$$\sin(\pi / p)\sin(\pi / r) \geq \cos(\pi / q).$$

Since the vertex figure $\{q, r\}$ must be one of the five Platonic solids

$$\{3, 3\}, \ \{3, 4\}, \ \{4, 3\}, \ \{3, 5\}, \ \{5, 3\},$$

the regular hypersolids can be enumerated by assigning these particular values to q, r, and using the inequality

$$p - 4 / p \leq 12 - 2q - r + 4 / r$$

or

$$p^2 - (12 - 2q - r + 4/r)p - 4 \leq 0 \tag{8}$$

to determine the possible values for p.

When $q = 3$ and $r = 3$, we have $p \leq (13 + \sqrt{313})/6 = 5.115 \ldots$.
When $q = 3$ and $r = 4$, we have $p \leq 4$.
When $q = 4$ and $r = 3$, we have $p \leq (7 + \sqrt{193})/6 = 3.48 \ldots$.
When $q = 3$ and $r = 5$, we have $p \leq (9 + \sqrt{481})/10 = 3.09 \ldots$.
When $q = 5$ and $r = 3$, we have $p \leq (1 + \sqrt{145})/6 = 2.17 \ldots$.

Thus the only finite polytopes $\{p, q, r\}$ are Schläfli's

$$\{3, 3, 3\},\ \{4, 3, 3\},\ \{5, 3, 3\},\ \{3, 3, 4\},\ \{3, 4, 3\},\ \{3, 3, 5\}.$$

A connection with group-theory is seen in Todd's observation that the direct symmetry-operations of the polytope $\{p, q, r\}$ constitute a group of order $(1/2)g$ for which the relations (2) provide an abstract definition. In fact,

$$R = R_1 R_2, \qquad S = R_2 R_3, \qquad T = R_3 R_4,$$

where R_1, R_2, R_3, R_4 are the reflections in the faces $P_1P_2P_3, P_2P_3P_0, P_3P_0P_1, P_0P_1P_2$ of the characteristic tetrahedron.

7. A Statistical Honeycomb

As we remarked in Section 3, the closest packing of equal circles in the Euclidean plane is provided by the incircles of the cells of the regular tessellation of hexagons $\{6, 3\}$. In fact, three equal circles are packed as closely as possible when they all touch one another, and the two-dimensional packing problem is easy because any number of further circles can be added in such a way as to continue the pattern systematically over the whole plane.

Analogously, four equal spheres are packed as closely as possible when they all touch one another, and some further spheres can be added so as to form the beginning of a pattern apparently consisting of the inspheres of the cells of a regular honeycomb $\{p, 3, 3\}$. This beginning can be continued for spheres of a suitable size in spherical (or elliptic) space with $p = 5$, and again in hyperbolic space with $p = 6$ (FEJES TÓTH, 1953, p. 159; COXETER, 1954, p. 266). The conclusion is inescapable that a compressed close-packing of equal lead shot, or a froth of equal bubbles, is trying to approximate to a Euclidean honeycomb $\{p, 3, 3\}$ in which p lies between 5 and 6. The fractional value of p means that this "honeycomb" exists only in a statistical sense, but the agreement with experiment is striking.

Setting $q = r = 3$ in the equation

$$p^2 - (12 - 2q - r + 4/r)p - 4 = 0 \tag{9}$$

(cf. (8)), we obtain

$$p^2 - (13/3)p - 4 = 0,$$

whence

$$p = \left(13 + \sqrt{313}\right)/6 = 5.115 \ \ldots,$$

in agreement with Matzke's observation that pentagons are prevalent (especially in froth) while hexagons are more frequent than quadrangles.

The cell $\{p, 3\}$ has an average of F faces, E edges, and V vertices, where, by (3),

$$F = \frac{12}{6-p} = \frac{23+\sqrt{313}}{3} = 13.56 \ \ldots,$$

$$E = \frac{6p}{6-p} = 17 + \sqrt{313} = 34.69 \ \ldots,$$

and

$$V = \frac{2}{3}E = 23.13 \ \ldots \ .$$

In conclusion, I wish to thank Michael Goldberg and John Satterly for drawing my attention to the experimental work of Matzke, whose estimate

$$F = 13.70$$

(see Section 4) motivated my choice of Eq. (9) in place of the equally plausible equation

$$\sin(\pi/p)\sin(\pi/r) = \cos(\pi/q),$$

from which the substitution $q = r = 3$ yields $p = \pi/\kappa$ in the notation of (COXETER, 1948, p. 293), whence

$$\frac{12}{6-p} = 13.398 \ldots, \qquad \frac{4p}{6-p} = 22.796 \ldots .$$

That other approach has been extended to n dimensions by ROGERS (1958).

Note added in proof. Professor BERNAL (1959) has used two independent experiments to obtain for F the approximate values 13.6 and 13.3. He refers to MEIJERING (1953, p. 282), who applied statistical methods of an entirely different kind to the so-called "Johnson-Mehl model", obtaining

$$V = 22.56.$$

REFERENCES

BARLOW, W. (1883) Probable nature of the internal symmetry of crystals, *Nature*, **29**, 186–188.

BERNAL, J. D. (1959) A geometrical approach to the structure of liquids, *Nature*, **183**, 141–147.

BOERDIJK, A. H. (1952) Some remarks concerning close-packing of equal spheres, *Philips Research Reports*, **7**, 303–313.

BRAHANA, H. R. (1928) Certain perfect groups generated by two operators of orders two and three, *Amer. J. Math.*, **50**, 345–356.

BURNSIDE, W. (1911) *Theory of Groups of Finite Order*, 2nd ed., Cambridge.

COXETER, H. S. M. (1948) *Regular Polytopes*, London.

COXETER, H. S. M. (1954) Arrangements of equal spheres in non-Euclidean spaces, *Acta Math. Acad. Sci. Hungaricae*, **5**, 263–274.

DEMPSTER, A. P. (1957) The minimum of a definite ternary quadratic form, *Canadian J. Math.*, **9**, 232–234.

DYCK, S. W. (1882) Gruppentheoretische Studien, *Math. Ann.*, **20**, 1–44.

FEJES TÓTH, L. (1953) *Lagerungen in der Ebene, auf der Kugel und im Raum*, Berlin.

FEJES TÓTH, L. (1953) On close-packings of spheres in spaces of constant curvature, *Publ. Math. Debrecen*, **3**, 158–167.

HALES, S. (1727) *Vegetable Staticks*, London.

Sir William R. HAMILTON (1856) Memorandum respecting a new system of roots of unity, *Phil. Mag.* (4), **12**, 446.

HILBERT, D. and COHN-VOSSEN, S. (1952) *Geometry and the Imagination* (translation of *Anschauliche Geometrie*), New York.

HULBARY, R. L. (1948) Three-dimensional cell shape in the tuberous roots of asparagus and in the leaf of rhoeo, *American Journal of Botany*, **33**, 558–566.

Lord KELVIN (=Sir W. THOMSON) (1887) On the division of space with maximum partitional area, *Phil. Mag.* (5), **24**, 503–514.

LEECH, J. (1956) The problem of the thirteen spheres, *Math. Gazette*, **40**, 22–23.

MARVIN, J. W. (1939) The shape of compressed lead shot and its relation to cell shape, *American Journal of Botany*, **26**, 280–288.

MATZKE, E. B. (1950) In the twinkling of an eye, *Bull Torrey Botanical Club*, **77**, 222–227.

MEIJERING, J. L. (1953) Interface area, edge length, and number of vertices in crystal aggregates with random nucleation, *Philips Research Reports*, **8**, 270–290.

MELMORE, S. (1947) Densest packing of equal spheres, *Nature*, **159**, 817; (1948) *Math. Rev.*, **9**, 53.

MILLER, G. A. (1935–59a) On the groups generated by two operators of orders two and three respectively whose product is of order six, *Collected Works*, University of Illinois, Urbana, Illinois, **2**, 107–110.

MILLER, G. A. (1935–59b) Groups defined by the orders of two generators and the order of their product, *Collected Works*, University of Illinois, Urbana, Illinois, **2**, 170–173.

MORDELL, L. J. (1948) The minimum of a definite ternary quadratic form, *J. London Math. Soc.*, **23**, 175–178.

PACIOLI (=PACCIOLI), L. (1509) *Divina Proportione*, Venice; (1946) Buenos Aires.

ROGERS, C. A. (1958) The packing of equal spheres, *Proc. London Math. Soc. (3)*, **8**, 609–620.

SCHLÄFLI, L. (1950) *Gesammelte Mathematische Abhandlungen*, vol. 1, Basel.

SCHÜTTE, K. and VAN DER WAERDEN, B. L. (1953) Das Problem der dreizehn Kugeln, *Math. Ann.*, **125**, 325–334.

STEINBERG, R. (1958) On the number of sides of a Petrie polygon., *Canadian J. Math.*, **10**, 220–221.

STEINBERG, R. (1959) Finite reflection groups, *Trans. Amer. Math. Soc.*

STEINHAUS, H. (1950) *Mathematical Snapshots*, 2nd ed., New York.

Sir D'Arcy W. THOMPSON (1952) *On Growth and Form*, 2nd ed., vol. 2, Cambridge.

TODD, J. A. (1931) The groups of symmetries of the regular polytopes, *Proc. Cambridge Philos. Soc.*, **27**, 212–231.

WEYL, H. (1952) *Symmetry*, Princeton.

Generalisations of the Kelvin Problem and Other Minimal Problems

R. PHELAN

Physics Department, Trinity College, Dublin, Ireland

Keywords: Dry Foam, Wet Foam, Structure, Minimal Problems

1. Introduction

In his 1887 paper Kelvin opens with the words "This problem is solved in foam". However his problem, that of finding a partitioning of space into cells of equal volume using the least possible surface area, is in fact a mathematical idealisation. The faces of the "foam" structure are assumed to be two-dimensional regions which intersect on one-dimensional curves. In reality a foam is a two-phase system, both phases occupying a nonzero volume. Following convention, for an aqueous foam we define ϕ_l to be the *liquid fraction*, i.e. the ratio of the liquid volume to the total volume.

The basis of Kelvin's analysis were the rules of equilibrium stated by PLATEAU (1873) for the dry foam ($\phi_l = 0$):

(a) The films which are the interfaces between cells have mean local curvatures, K, related to the pressures in the adjoining cells by Laplace's law.

$$\Delta p = 2\sigma K$$

(b) The films meet three at a time on lines.

(c) The lines meet four at a time at vertices.

(d) The local symmetry at these lines is threefold, and the vertices are tetrahedrally symmetric.

These rules follow easily from arguments of surface tension, although a full proof of (b) and (c) with mathematical rigour was given only after a century (TAYLOR, 1976; ALMGREN and TAYLOR, 1976, 1996).

The problem can be generalised, and indeed real foams demand this because of their finite liquid content. We will here consider this generalised problem and other analogous problems from further afield. Before doing so, we review the present state of our understanding of the variation of energy(area) among the various structures which obey Plateau's rules.

(a)

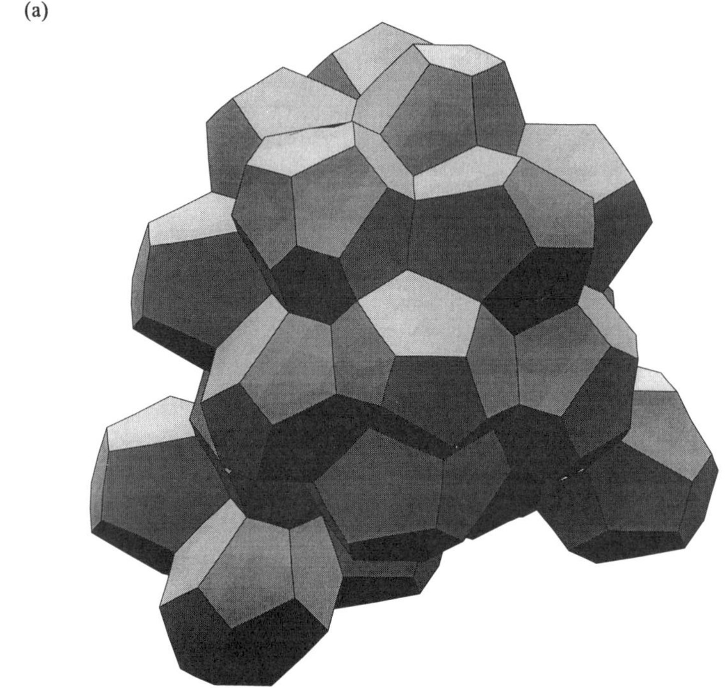

(b)

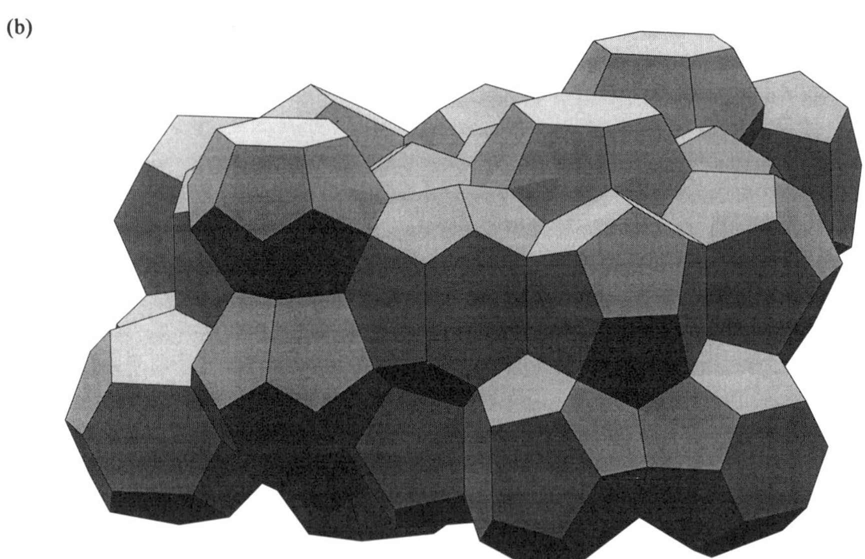

Fig. 1. Other examples of possible foam structures drawn from the same class of chemical structures as A15. (a) C15, (b) σ (β-Uranium).

2. Dry Foam

So far, no rival structure has been found to better that reported by WEAIRE and PHELAN (1994a). While unsuccessful this search has focused attention on a whole class of closely related structures. First described and categorised in the context of crystallography (SHOEMAKER and SHOEMAKER, 1986) these crystalline phases define polyhedral geometries consisting of only tetrahedral vertices. In the notation of this subject A15 underlies the WP structure. Figure 1 shows two further ordered foams derived from the C15 and σ (or β-Uranium) phases respectively. The energy variation among these structures invites interpretation in terms of their topological parameters (RIVIER, 1994; RIVIER and ASTE, 1995, 1996). Since $\bar{z}$, the average number of faces per cell, is often considered to be an important variable in any such description we present a plot of energy per unit volume for several structures versus $\bar{z}$ in Fig. 2, several of these were constructed

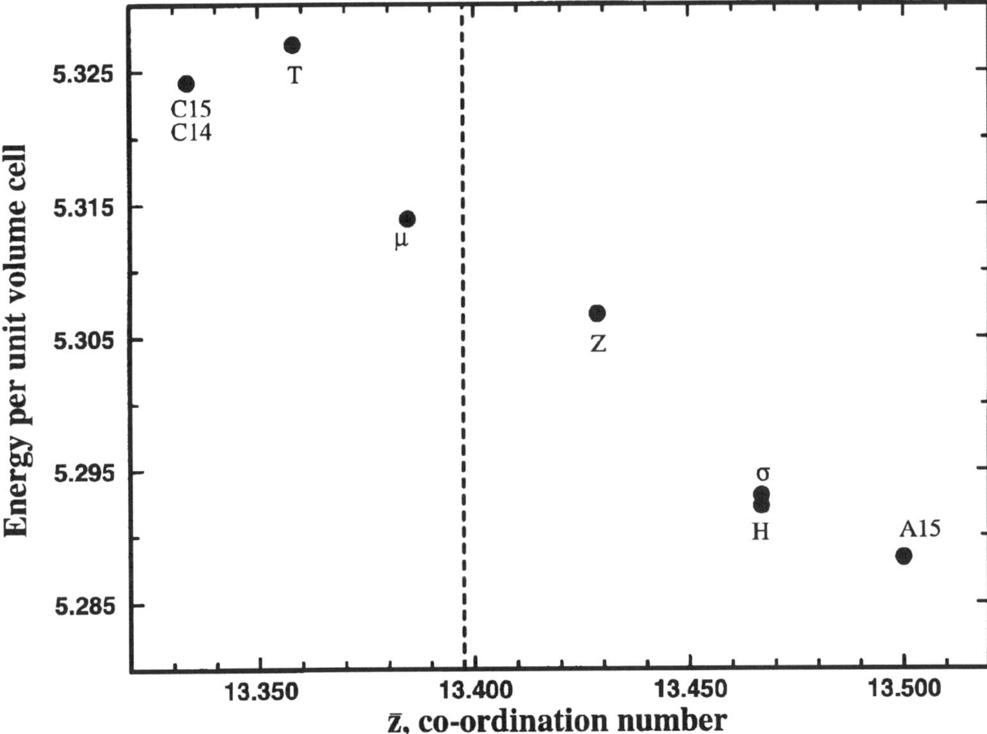

Fig. 2. The energy per unit volume of unit volume cells for several of the structures listed by SHOEMAKER and SHOEMAKER (1986) plotted against the average co-ordination number, $\bar{z}$. Not included is the Kelvin structure at an energy of $\approx$5.3063 at $\bar{z} = 14$. The vertical dashed line indicates the value $\bar{z} \approx 13.397...$ often considered to be ideal.

by J. Sullivan, R. Kusner and A. Kraynik (KRAYNIK, 1996; KUSNER and SULLIVAN, 1996; Kraynik, private communication, 1995; Sullivan, private communication, 1995).

Various arguments have often suggested the value $\bar{z} \approx 13.397$ to be optimal in this problem (COXETER, 1996). OGEÉ (1994) has calculated the energy for a hypothetical structure with this co-ordination to be ≈5.2544. This is based on a single cell type of $12/(6 - n) \approx 13.397$ faces where n, the number of edges per face, is equal to $\frac{2\pi}{\cos^{-1}(1/3)} \approx 5.104$. These values arise from the geometry of a regular tetrahedron. Such a structure cannot, of course, exist but one can use non-integral values in various formulae to arrive at this interesting number.

3. The Wet Foam Problem

For a physical foam the faces are in fact two closely separated (air-liquid) interfaces, for example a liquid thin film separating two regions of gas. The 'lines' of intersection of these films are veins of liquid meeting at thickened vertices. Clearly these considerations affect the nature of the minimisation problem. An equilibrium foam of macroscopic cells (of the order of several millimetres) contains only a small volume of liquid. Fine foams or systems such as microemulsions have equilibrium states of much higher liquid fraction. Even for a macroscopic foam it is quite easy, using the forced drainage procedure of WEAIRE *et al.* (1993), to observe foams of high liquid fraction in quasi-equilibrium. For such differing systems we do not expect to observe the same optimal structures.

Let us define more precisely, the mathematical problem posed by such wet foams, as a generalisation of Kelvin's problem.

Problem: *Consider a finite region of volume V. We wish to embed N, equal volume, non-intersecting regions of total volume $V_l < V$ with minimum possible surface area. In the limit $V \to \infty$, $V_1 \to \infty$ and $N \to \infty$ such that V_1/V and V_1/N remain constant we recover the analogue of the global problem posed by Kelvin.*

In the wet limit the bubbles become spherical, since the sphere is the best local area minimiser, and the problem reduces to that of finding a densest sphere packing. The best such structure is a close-packed arrangement of hexagonal layers, e.g. fcc (although this apparently remains to be proved!). Obviously the topology of this structure is very different from that of either A15 or bcc. For liquid fractions between 0 and about 0.26 (the void volume of a sphere close packing), it is unclear which structures should prevail, and what should be the nature of the transitions between them. These questions were first posed by WEAIRE (1994). Figure 3 shows the form of the conjectured energy dependence on liquid fraction for fcc and bcc with the limits of stability for each structure indicated. Kelvin's structure (bcc) should remain stable as the liquid fraction is increased from zero until the quadrilateral contacts are lost. While for fcc, as the liquid fraction is reduced from the limiting value of ≈0.26 there should be a loss of stability corresponding to the decay of the eight-fold vertices to lower energy four-fold ones. This scenario has required revision in some respects as explained below.

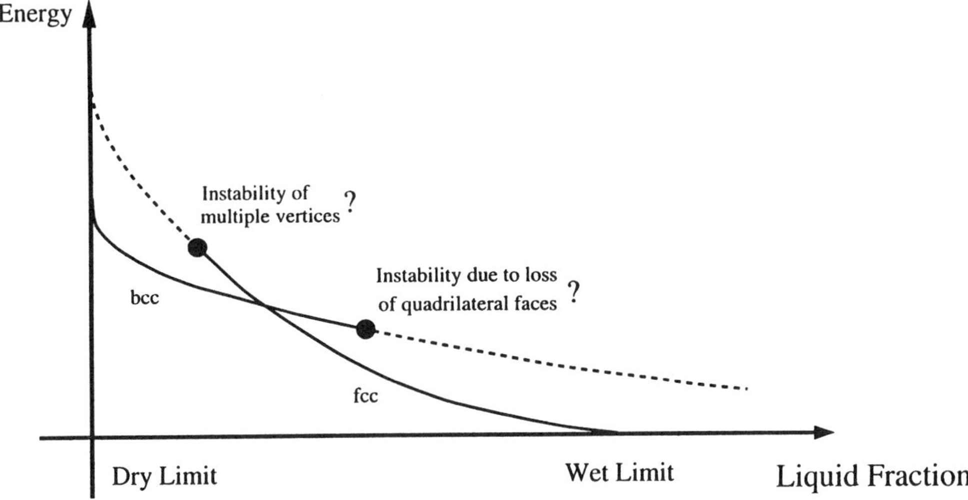

Fig. 3. Highly schematic diagram of the energy variation with liquid fraction of the fcc and bcc (Kelvin) structures, as suggested by WEAIRE (1994).

4. Calculations for Wet Foams

Calculations have been made on wet foam structures, by use of an adapted version of the Surface Evolver (BRAKKE, 1992; PHELAN *et al.*, 1995). The idealised edges and vertices of the dry foam (e.g. see Fig. 1) are replaced by liquid regions (Plateau borders) of the form show in Fig. 4.

Applying this procedure to the Kelvin and fcc structures yields the corresponding wet foam structures shown in Fig. 5. We are interested in exploring the structural transitions between foam topologies as the liquid fraction is increased. Changing the relative volume of the Plateau borders we have calculated the energies of these rival structures as a function of liquid fraction, ϕ_l. Figure 6 shows this variation for Kelvin's structure fitted to the expansion, suggested by WEAIRE (1994),

$$E = E_d + E_l \phi_l^{\frac{1}{2}} + E_v \phi_l + \cdots \qquad (2)$$

where the first term is the energy of the dry system, the second is the correction due to thickening of the lines and the third makes a final correction for the thickening of the vertex. Note the high quality of the fit suggesting a simple interpretation of the detailed geometry of the Plateau borders.

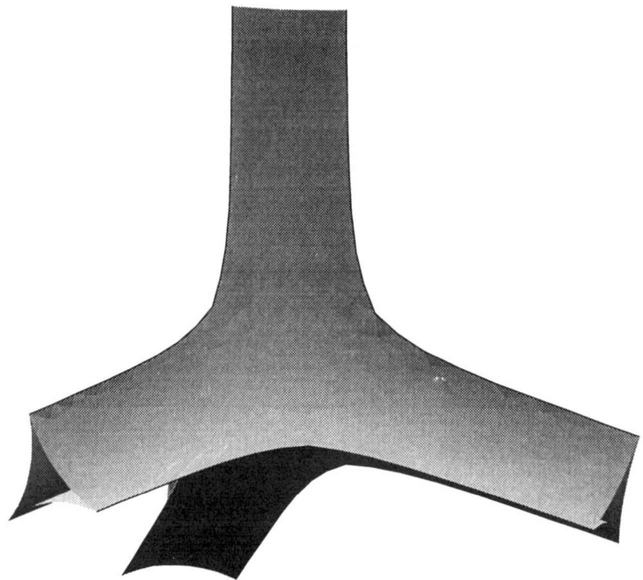

Fig. 4. Detail of a single wet foam junction formed by four Plateau borders.

This work has provoked another surprising result unanticipated by WEAIRE (1994). It arises from a reconsideration of Plateau's vertex rule (c) as discussed above. The first clue to this was the finding that fcc does *not* spontaneously undergo any vertex instability in simulations, as the liquid fraction is decreased. That is, the instability suggested for fcc in Fig. 3 is never encountered in such calculations.

The conclusion of a detailed analysis is that an eight-fold symmetric vertex is stable for any finite liquid fraction (WEAIRE and PHELAN, 1996). In a sense this does not contradict Plateau's rule, since it is stated for a dry foam, but all real foams are wet, to some extent. The range of distortion over which this multiple vertex (or fcc) is stable is limited however and of the order of the Plateau border width. Hence as the liquid fraction tends to zero, smaller and smaller distortions will provoke the decomposition to four fourfold vertices.

In contrast, there is a well defined structural transition in the case of bcc (Fig. 5(a)). As the liquid content is increased from zero, the quadrilateral faces shrink and eventually disappear at a liquid fraction of $\approx 11\%$ in remarkable agreement with the early crude estimate of WEAIRE *et al.* (1993). This topological change induces an elastic instability and a structural transformation.

(a)

(b)

Fig. 5. (a) Kelvin's structure for a non-zero liquid fraction. (b) The cubic unit cell of the fcc wet foam structure. Note the obvious presence of eight-fold vertices.

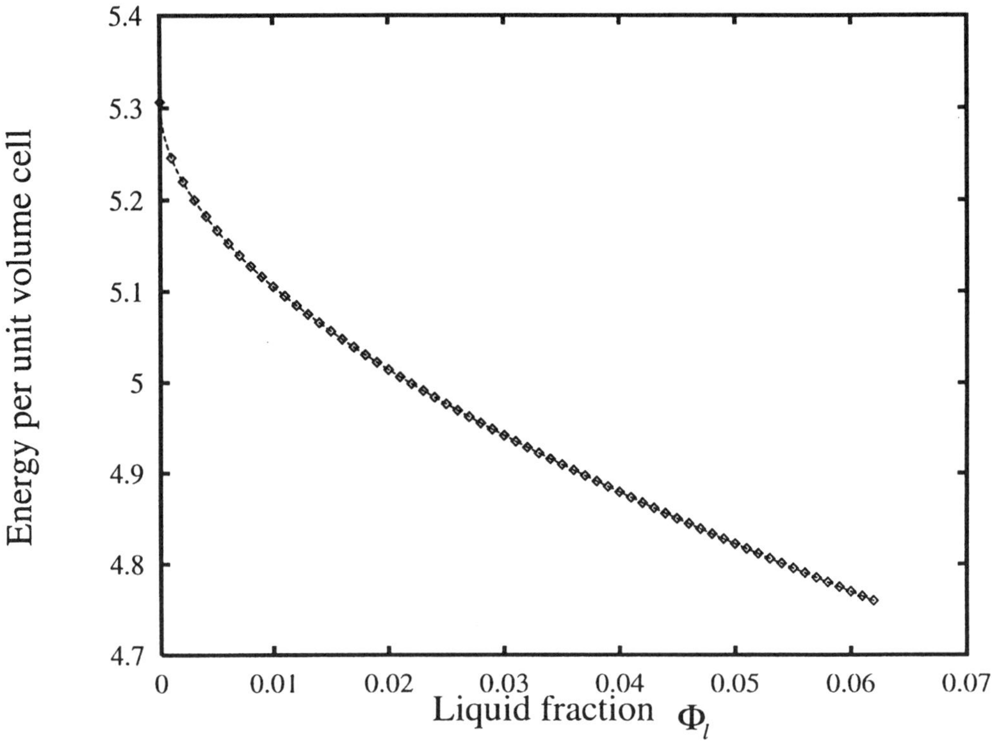

Fig. 6. The energy per unit volume for Kelvin's structure as a function of liquid fraction fitted to the expansion $E = E_d + E_l \phi_l^{1/2} + E_v \phi_l$ as described in the text.

5. Other Dimensions

The Kelvin problem is posed in three dimensions, one can of course pose it in any number. It is trivial in one, in two dimensions it is related to the problem of the bee's honeycomb. We are not aware on any published work in higher dimensions, but one can foresee an attack on the general problem before long.

5.1. The honeycomb problem

The honeycomb has been celebrated for many centuries as the epitome of economy in nature (Fig. 7). It divides the two-dimensional plane into cells of equal area with minimum line length, or so it is assumed. It has been applied (or misapplied) to the explanation of many natural phenomena (WEAIRE and RIVIER, 1984).

There does not seem to be any proof of this generally accepted minimal isoperimetric property of the honeycomb partitioning. There is however a weaker theorem which states

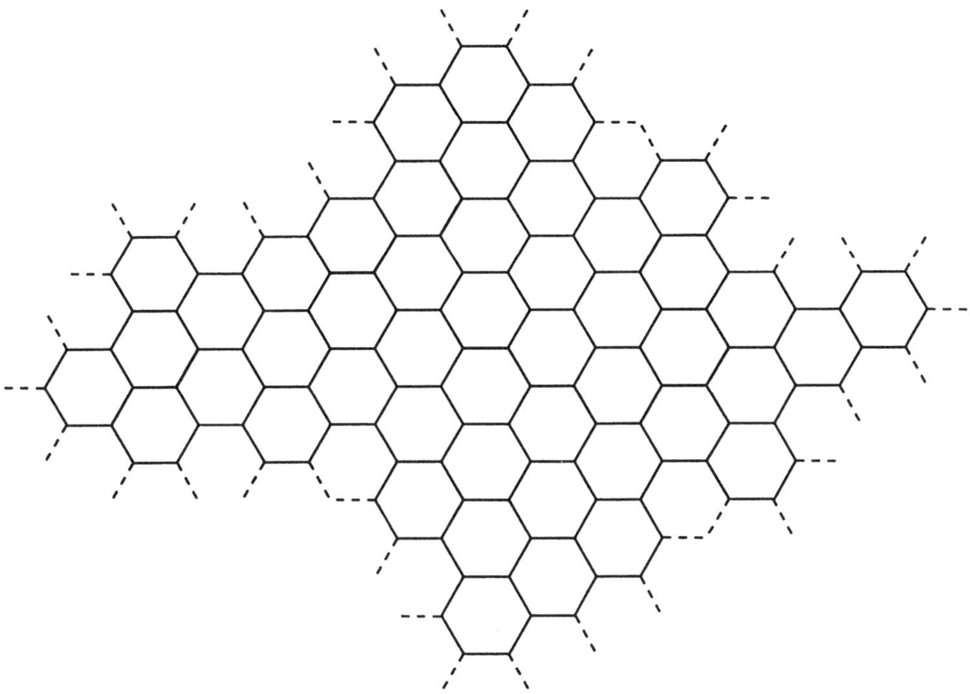

Fig. 7. The regular hexagonal tessellation ("honeycomb partitioning").

that of all tessellations consisting of *convex* cells of equal area this partitioning yields the least possible total edge length due to FEJES TÓTH (1964). The proof is involved and perhaps unintuitive. He also remarks that if one relaxes the constraint of convexity the "proof seems to involve considerable difficulties". A convex polygon is one for which any two interior points may be joined by a line segment lying wholly in the interior. For the present problem, convexity forces the cells to be straight edged polygons.

Here we offer an alternative analysis of the problem (Phelan and Weaire, unpublished). We consider only straight edged polygonal cells but do not require convexity. We also consider only trivalent vertices but this entails no loss of generality because any vertex of more that 3 edges can be redefined as two or more degenerate trivalent vertices, joined by a side of zero length as in Fig. 8. The requirement that all edges be line segments is quite severe. In general, circular arcs are consistent with the minimal condition, straight edges being just the special case of infinite radius.

5.2. *Analysis*

We wish to progress towards a proof of the following result:

Theorem 1: *Among all partitions of the plane into polygons of specified equal area, the honeycomb partition attains the least possible edge-length per cell.*

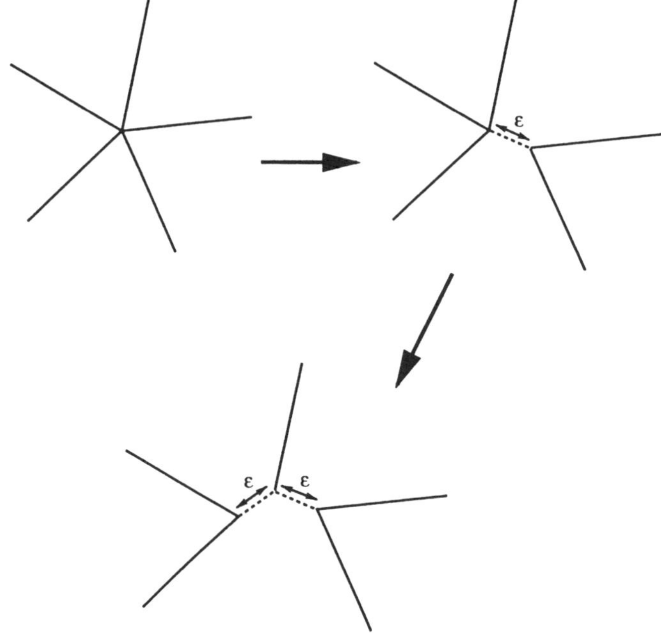

Fig. 8. Example of a multiple vertex and its equivalent decomposition into trivalent vertices connected by edges of length $\varepsilon \to 0$.

Let us take the case of cells of unit area. We consider all such structures, characterised in part by the topological distribution function $f(n)$. Although a consistent definition of $f(n)$ presents difficulties (Kusner, private communication, 1995; Brakke, private communication, 1995) and one can imagine pathological examples, it gives in principle the relative proportions of n-sided cells. In particular the honeycomb structure corresponds to

$$f(6)=1$$

$$f(n)=0, \qquad n \neq 6. \tag{3}$$

Any structure of the kind considered must satisfy:

$$\sum_n f(n)=1 \qquad \text{(normalisation)} \tag{4}$$

$$\sum_n (n-6)f(n)=0 \qquad \text{(Euler's Theorem).} \tag{5}$$

The latter plays a crucial role in what is to follow. There is no ambiguity in the relations (3), (4) and (5) provided $f(n) = 0$ only when the number of n-gons = 0. For example, in a finite structure or an infinite structure generated by copies of some finite, fundamental region. However, for a general infinite structure, finite numbers of n-gons may be added without changing the $f(n)$ values. We cannot distinguish between structures related in this way.

It is an established result that the minimal n-gon of given area is regular (see for example FEJES TÓTH, 1964). An elementary proof can be constructed by first proving that the edge lengths must be equal and then that the angles must be equal, simply by considering local adjustments of these quantities. Let the perimeter of the minimal n-gon be $\hat{L}_n$. This is a monotonic decreasing and concave function of n. For any given structure the average perimeter is

$$L = \frac{1}{2}\sum_n f(n) L_n \tag{6}$$

where L_n is the mean perimeter of the n-gons in the structure. Clearly,

$$L_n \geq \hat{L}_n. \tag{7}$$

The factor of 1/2 in (6) comes from the fact that in such a structure each edge is shared by two polygons.

Because of the relation (5), we can subtract from L_n any function of the form $a(n-6)$, where a is a constant, without altering the value of L for the structure. We choose a, (e.g. $a = -0.07$), such that $L_n - a(n-6)$ has its minimum at $n = 6$. This can be achieved, since L_n is concave. Clearly the minimum value is just L_6. For convenience we now subtract L_6 from each point to obtain the plot shown in Fig. 9 where we have defined the line length adjusted by the transformations described above as,

$$\lambda_n = L_n - \left(a(n-6) + b\right) \tag{8}$$

with $a = -0.07$ and $b = L_6$.

Now recall that the minimum solution must satisfy (4). Clearly L is bounded below by $(1/2)\hat{L}_6$. Since this value can be realised (in the symmetric honeycomb topology) it must be the global minimum of L. We have not shown that this is the unique global minimum. For example, introducing a finite number of defects does not affect the relation (3) and does not change the average perimeter defined in (6).

Noting the caveat regarding $f(n)$ mentioned above, this analysis represents a concise proof that the honeycomb partitioning attains the least edge-length density of all partitionings of the plane into cells of equal volume with the constraint that all edges are

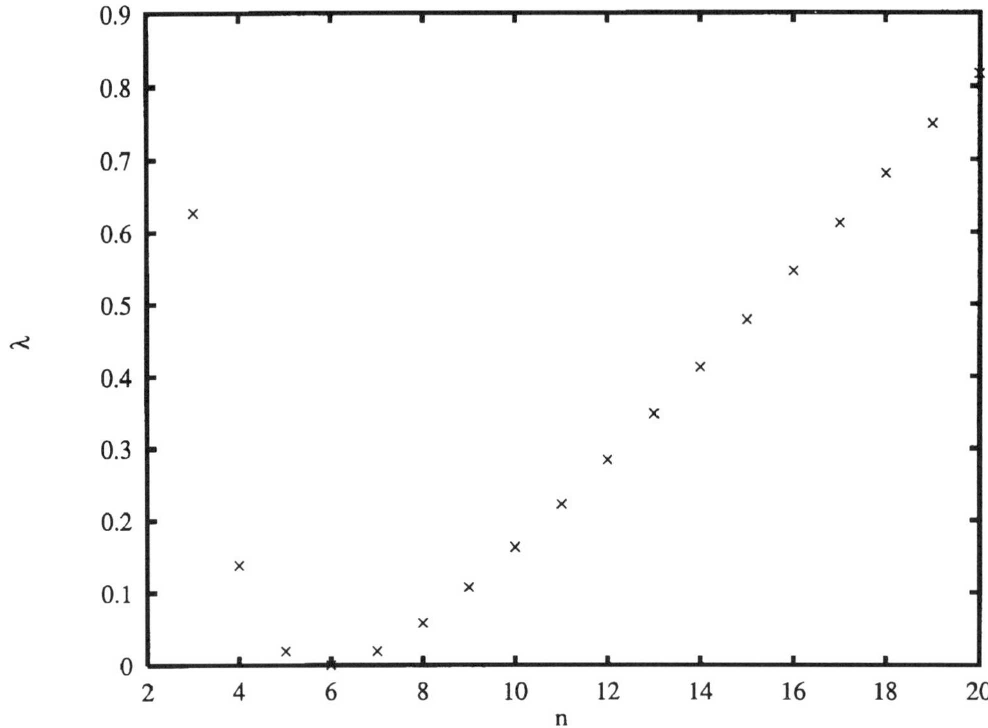

Fig. 9. The adjusted line length, λ_n, obtained by subtracting $-0.07(n-6)+L_6$ from the line length L_n.

straight line segments. Comment from the mathematical community (Morgan, private communication, 1996) has indicated that FEJES TÓTH (1953) may have used similar arguments. The general case still remains to be addressed.

6. Confined Structures: Cylindrical Structures, Slabs and Surfaces

Returning now to three dimensions; what of experiment? The study of bulk ordered structures is frustrated by the failure of experimental samples to order, although it is true that small fragments of the WP structure have been observed (WEAIRE and PHELAN, 1994b, reprinted this volume).

This difficulty is not encountered when relatively large bubbles are introduced into a cylindrical tube. Indeed any one of a remarkable variety of beautiful ordered structures can be created in a few seconds, depending on the ratio of tube to bubble diameter. They will surely provide the ideal test-bed for future experiments of many kinds. In the standard notation of the subject (PITTET *et al.*, 1995), Fig. 10 shows Evolver models of the 211 and 422 cylindrical structures.

Ordered structures also form readily at plane surfaces and in slab geometries, as

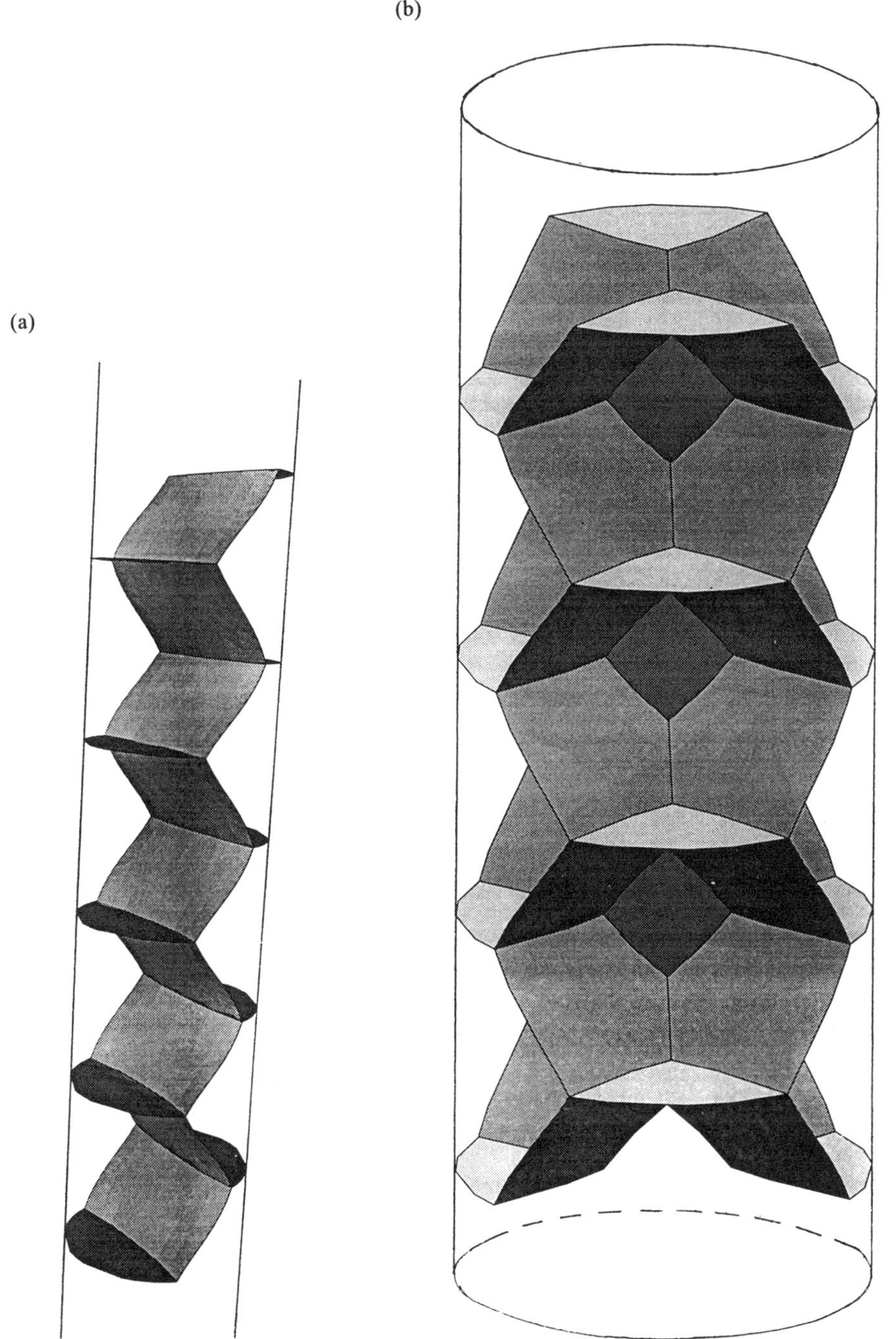

Fig. 10. Two simple arrangements of foam cells in a cylinder. (a) the 211 and (b) the 422 structures.

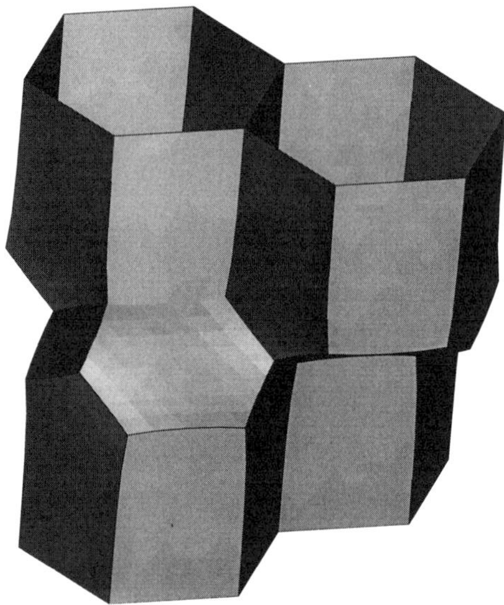

Fig. 11. Four Tóth cells, described by him in the context of the bees' honeycomb and also observed as a typical foam surface cell.

discussed by WEAIRE and PHELAN (1994b, reprinted here). Figure 11 shows a calculation of a double layer of foam cells between two glass plates of the type described by Fejes Tóth in the context of the bees' honeycomb and easily realised experimentally for a foam. (FEJES TÓTH, 1964; WEAIRE and PHELAN, 1994b, 1994c). If the liquid fraction is increased there in a structural transition to a close packed topology.

7. The Wigner Crystal and Other Related Problems

So far we have restricted our attention to the isoperimetric problem dictating foam structure. Many other mathematical and physical problems exist for which the topologies, particularly fcc and bcc, discussed above occur as solutions (or more usually as best known solutions).

We have already mentioned optimal sphere packings, i.e. fcc and related structures, in the context of the wet foam limit. The packing density for fcc is 0.74048 and 0.68017 for bcc. There is a corresponding *covering* problem where the spheres are allowed to overlap and one must completely fill space as efficiently as possible. For this question, bcc is the best know arrangement with a covering density of 1.4635 compared to 2.0944 for fcc. (CONWAY and SLOANE, 1993).

Another interesting example is that of constructing digital representations of a continuous parameter space. We wish to quantise a continuous region into a discrete set

Table 1. The Ewald energy and the energy per unit volume cell of a dry foam for various structures. The lower bounds correspond to a single unit volume sphere in the case of a foam and an isolated point charge in a single uniform atomic sphere in the case of the Ewald energy.

Structure	Ewald energy	Dry foam energy
bcc	–1.791859	5.30628
fcc	–1.791747	5.34539
A15	–1.787680	5.28834
C15	–1.772291	5.32421
Simple cubic	–1.760119	6.00000
Lower bound	–1.800000	4.83598

of representative values. The quantity to be minimised is the error that this introduces (measured as the square of the distance from the original point to its quantised representation). For example, in one dimension a set of equally spaced points on the line is the best quantiser and this is the representation scheme used in digital computers. The best known three dimensional quantiser is the Voronoi structure based on the bcc lattice. In fact the Voronoi construction plays an important part in the analysis of many of these mathematical problems (CONWAY and SLOANE, 1993).

Questions of a similar kind occur in physics also. That of the so called Wigner crystal (or jellium) is an example. As an aid to understanding the crystal structure of metals it is useful to consider the following model. One replaces the atomic positions by point positive charges and the surrounding electrons by a uniform negative background such that the whole system is neutral. The electrostatic energy (Ewald energy) of such systems can be calculated numerically (e.g. HARRISON, 1966). The lowest known Ewald energy is that of the bcc lattice (see Table 1). We have calculated the energy for the C15 and A15 structures in the hope of improving on this result, however as Table 1 shows neither structure is particularly efficient. In fact while fcc is a poor dry foam it is a better Ewald minimiser than either C15 or A15. In contrast a simple cubic structure is inefficient in both cases. The lower bound, $E_{\mathrm{Ewald}} = -1.8$, for the Ewald energy has been proved to be that of an isolated point charge in a uniform atomic sphere (LIEB and NARNHOFER, 1975).

The geometric properties of the sphere are a common thread running though all of these questions. The frustration between its failure to fill space and its various minimal properties is of central importance. However the correlation and variation between the problems considered remains to be explored fully. The phases which include A15 and which have proved of such significance in the study of foam structure will provide an interesting test bed for this exploration.

Note added in proof. Brakke (private communication) has found negative Hessian eigenvalues (indicative of instability) in direct calculations of higher-order vertices using the Evolver. Concluding in particular that the eight-fold symmetric vertex discussed here is *unstable* below a liquid fraction of 0.00027.

This work was supported by the FOAMPHYS Network, Contract ERBCHRXCT940542, Shell Research and a Forbairt Basic Research Award. We would like to thank Ken Brakke, John Sullivan, Rob Kusner, Andy Kraynik, Nicolas Rivier, Frank Morgan and Aladár Heppes for many helpful discussions and John Sullivan for supplying us with Evolver definitions of many of the dry foam structures discussed above. RP would like to thank Rob Kusner for an invitation to visit the GANG lab at UMass, Amherst during the summer of 1995 and Stuart Findlay for collaboration on some of the calculations.

REFERENCES

ALMGREN, F. J. and TAYLOR, J. E. (1976) *Scientific American*, July, pp 82–83.

ALMGREN, F. J. and TAYLOR, J. E. (1996) *Forma*, **11**, 199–207.

BRAKKE, K. (1992) *Exp. Math.*, **1**, 141.

CONWAY, J. H. and SLOANE, N. J. A. (1993) *Sphere Packings, Lattices and Groups*, A series of comprehensive studies in mathematics, Springer-Verlag, New York.

COXETER, H. S. M. (1996) *Forma*, **11**, 271–285. Originally published: (1958) *Illinois Journal of Math.*, **2**, 746.

FEJES TÓTH, L. (1953) *Lagerungen in der Ebene auf der Kugel und im Raum*, Die Grundlagen der Mathematischen Wissenschaften in Einzeldarstellung, Bd. 65, Springer, Berlin.

FEJES TÓTH, L. (1964) *Regular Figures*, International Series of Monographs on Pure and Applied Mathematics, Pergamon Press, Oxford.

HARRISON, W. A. (1966) *Pseudopotentials in the Theory of Metals* (ed. W. A. Benjamin), New York.

KRAYNIK, A. (1996) *Forma*, **11**, 255–270.

KUSNER, R. and SULLIVAN, J. (1996) *Forma*, **11**, 233–242.

LIEB, H. and NARNHOFER, H. (1975) *J. Stat. Phys.*, **4**, 291.

OGEÉ, J. (1994) unpublished T.C.D. report.

PHELAN, R., WEAIRE, D. and BRAKKE, K. (1995) *Exp. Math.*, **4**, 181.

PITTET, N., RIVIER, N. and WEAIRE, D. (1995) *Forma*, **10**, 65.

PLATEAU, J. A. F. (1873) *Statique Expérimentale et Théorique des Liquides Soumis aux Seules Forces Molécularies*, Gauthier-Villars, Paris.

RIVIER, N. (1994) *Phil. Mag. Lett.*, **69**, 297.

RIVIER, N. and ASTE, T. (1995) *Phil. Trans. Roy. Soc.*, **304**, 2055.

RIVIER, N. and ASTE, T. (1996) *Forma*, **11**, 223–231.

SHOEMAKER, D. P. and SHOEMAKER, C. B. (1986) *Acta Crystallogr. B*, **42**, 3.

TAYLOR, J. E. (1976) *Ann. Math.*, **103**, 489.

THOMPSON, W. (Lord KELVIN) (1887) *Phil. Mag.*, **24**, 503.

WEAIRE, D. (1994) *Phil. Mag. Lett.*, **69**, 99.

WEAIRE, D. and PHELAN, R. (1994a) *Phil. Mag. Lett.*, **69**, 107, included in this volume.

WEAIRE, D. and PHELAN, R. (1994b) *Phil. Mag. Lett.*, **70**, 345, included in this volume.

WEAIRE, D. and PHELAN, R. (1994c) *Nature*, **367**, 123.

WEAIRE, D. and PHELAN, R. (1996) *J. Phys.: Condens. Matter*, **8**, L37.

WEAIRE, D. and RIVIER, N. (1984) *Contemp. Phys.*, **25**, 5.

WEAIRE, D., HUTZLER, S. and PITTET, N. (1992) *Forma*, **7**, 259.

WEAIRE, D., PITTET, N., HUTZLER, S. and PARDAL, D. (1993) *Phys. Rev. Lett.*, **71**, 2670.

Establishment of Epidermal Cell Columns in Mammalian Skin: Computer Simulation*

H. HONDA[1]**, T. MORITA[2] and A. TANABE[2]***

[1]*Kanebo Institute for Cancer Research, Misakicho 1-9-1, Kobe 652, Japan*
[2]*Laboratory of Experimental Radiology, Aichi Cancer Center Research Institute, Tashirocho, Nagoya 464, Japan*

Abstract. The present investigation shows an example of tissue formation in relation to individual cell properties. Epidermis of mammalian skin has recently been shown to be organized during ontogeny into neat vertical cell columns, in which a cell approximates the flattened form of Kelvin's tetrakaidecahedron, a 14-sided body with eight hexagonal faces and six square faces. Such an epidermal architecture is compatible with the organization and turnover of stacked cells, a constant loss of surface cells and a supply that cells from a basal layer differentiate during migration to the surface. The developmental process from irregular cell aggregate into neat columns in skin is simulated on a digital electronic computer with the assumption as follows: a new cell migrating upward from a basal layer jostles and settles at the less crowded area among upper cells. After migration of many cells to the surface, the simulation demonstrates that cells have been stacked in neat columns, and the stability of the architecture is also shown.

1. Introduction

This is a presentation of an example of tissue formation in relation to individual cell properties. Organization of mammalian skin into remarkable ordered structure has been simulated on a digital electronic computer with the simple assumption of individual cell properties.

In addition to an observation that the surface of mammalian skin is predominantly covered by hexagonally shaped squamous corneocytes, several researchers have recently discovered that the stratum corneum and the superficial living epidermis of the skin is organized into neat vertical cell columns (e.g. Fig. 1 in MACKENZIE, 1969; Fig. 1 in

---

*Reproduced from *Journal of Theoretical Biology* (1979) Vol. **81**, 745–759.

Keywords: Computer Simulation, Dirichlet, Epidermal Cell, Kelvin's Tetrakaidecahedron, Voronoi (given by the editors of FORMA).

**Author for correspondence, current address: Hyogo University, Kakogawa, Hyogo 675-01, Japan.

***School of Dentistry, Aichigakuin University, Nagoya 464, Japan.

CHRISTOPHERS, 1972; Fig. 6 in MENTON, 1976a). These studies have been accomplished instead of using histological processing which severely distorts the true structure, by treating the skin with an alkaline solution which swells epidermis and leaves the cell membranes intact (MACKENZIE, 1969), and by staining cell boundaries with a fluorescent dye (CHRISTOPHERS, 1971) or silver nitrate (MACKENZIE and LINDEN, 1973).

Most of mammalian epidermis consist of tightly packed columns of squamous cells stacked normal to the surface of the skin. Epidermis suffers a constant loss of surface cells and is supplied with cells from a basal cell layer. The cells from a basal layer are differentiated into spinous, granular and cornified cells successively during migration to the surface (CHRISTOPHERS *et al.*, 1974; POTTEN and ALLEN, 1975; ALLEN and POTTEN, 1976). MENTON (1976a) has performed a detailed observation of mammalian epidermis, compared it with cork cambium and the pith of woody plant stems, and found that the shape of a cell in epidermis approximates closely the shape of a tetrakaidecahedron flattened normal to the surface of the skin (Fig. 1). An orthic tetrakaidecahedron is a 14-sided body with eight hexagonal faces and six square faces (all the edges of the faces are equal), and represents Lord Kelvin's mathematical solution for division of space, without interstices, into uniform bodies of equal volume and minimal surface (e.g. THOMPSON, 1942). The geometrical assumption that epidermis consists of flattened tetrakaidecahedra is compatible with the organization and the dynamical process of cell loss at surface and a supply from a basal layer. Other researchers supported this assumption (ALLEN and POTTEN, 1976).

On the other hand, there is a finding that cell arrangement in the dorsal epidermis of a mouse is irregular in the early stages of development. There are no definite columns of epidermal cells. Cell boundaries overlap irregularly through every focus level in a transparent specimen. During ontogeny the neat vertical cell columns in epidermis become to be established from irregular cell aggregates (unpub. data by T.M. and A.T.).

The present report demonstrates that this developmental process can be simulated by

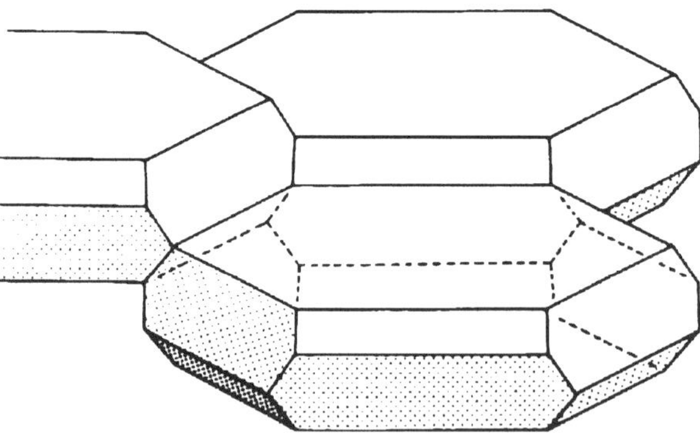

Fig. 1. Flattened tetrakaidecahedra neighboring each other are not on the same layer and have eight hexagonal faces (two of them are large and the remaining six are small) and six square faces respectively.

computer with an assumption that a cell migrating upward from a basal layer jostles and settles at the less crowded area among the upper cells. The stability of the ordered architecture when it is once established is also investigated.

2. Methods

Computer programs are written with the assembler language. The program of the two-dimensional Dirichlet domains which is used for approximate two-dimensional representation of the results is the same as described in HONDA (1978). The computations are carried out on a digital electronic microcomputer and a disc memory (P652 and DAS 604, Olivetti) and the results are drawn by an *X*-*Y* plotter (WX535, Watanabe Sokuki, Tokyo).

The distribution of the points of the 0th layer, which is irregular but each point has six neighbors, is made as follows: many short paper columns which stand positioned on the smooth surface, are packed loosely by a string lariat (lasso) but not in the closest packing. The *x*- and *y*-coordinates of the column centers are recorded and used on a digital computer.

The periodic boundary condition of the rectangle ($2:\sqrt{3}$) or the rhomb (60°) is used for computation. Both give similar results. In the case of the periodic boundary condition of the square deformed hexagons are finally obtained.

3. Results

A cell aggregate could be approximately described by a point distribution using the geometrical concept of Dirichlet domains (HONDA, 1978); a Dirichlet domain (or Voronoi polyhedron) represents an effective area which is influenced from a corresponding point. It is based on the fact that a cell in an aggregate must establish its boundary against contiguous cells. Several examples of two-dimensional cellular patterns have been shown in HONDA (1978). In the case of mammalian skin, the cell in the ordered structure of epidermis is flattened normal to the skin surface, and the ratio of diameter to thickness is 10~15 even when specimens are swollen by alkaline treatment. We assume in such a system that a point (which represents a cell from a basal layer) is placed on a plane according to the positions of upper points (which represent upper cells). We are interested in how the distribution of the points is determined in a plane parallel to skin surface. It is not the purpose of the present investigation how the thickness of a cell (or the vertical distance between points) is determined. (The thickness may be related to the surface tension and longitudinal compression of the skin tissue owing to the expansion of a mammalian body.) The vertical distance between points is assumed to be a small constant.

We will simply consider the two-dimensional case at first. The points are randomly distributed on the uppermost line (solid circles in Fig. 2) and points on lower lines are positioned by the upper points; a point on the lower line is determined to be midway between the nearest upper points. All points are arranged after such manner. When the processes are sequentially performed from the upper line to the lower, the distribution of

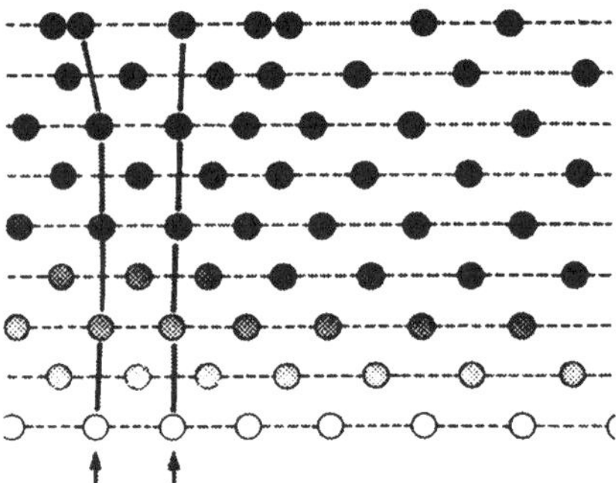

Fig. 2. Explanation of mechanism of cell-column-formation in the two-dimensional space. A random distribution of points on the upper line becomes uniform during piling downwards. The points on every other layer are approximately arranged on a vertical line (arrows). A point on the lower line is determined to be midway between the nearest upper two points. Periodic boundary condition is used for the points next to the terminals.

the points becomes uniform (open circles in Fig. 2). The points on every other layer are approximately arranged on a vertical line (arrows in Fig. 2). The similar process can be performed in the actual three-dimensional space; a point from a basal layer migrates upward and is positioned by the upper three points which are on a plane approximately parallel to the skin surface. The point is determined to be just below the central part of the triangle consisting of the upper three points, instead of midway between the nearest upper two points.

Hereafter, we will often use the concept of a *layer* to which points (cells) belong. By way of caution we have to indicate that a layer is not material substance and is defined by connection among neighboring points parallel to the skin surface. The cells which belong to *one* layer do not make a continuous surface by themselves, but the cells which belong to three sequential layers make a surface as is shown later (Fig. 6). A layer is tentatively assumed to be flat for simplicity.

For initiation of the computer simulation in the three-dimensional space, we have to make an initial condition corresponding to an irregular cell aggregate. Point-positions on only two layers (the 0th and 1st) are enough for the computer simulation, which are the innermost (lowest) surface of the cell aggregate of epidermis in the early stages of development. As shown in Fig. 3(a), which is a view from below (cells are assumed not to be spherical, but to be space filling polyhedra positioned at the centers of the spheres), the points are irregularly distributed on the 0th layer (the uppermost layer in the computer simulation), but so that each of them has six neighbors, as described in Methods. (cf. In the case of the closest packing of spheres in a plane each sphere has *six* neighbors.) By connecting neighboring pairs of the points with lines, we get a network consisting of triangles which cover a plane (Fig. 3(a)). The points on the 1st layer which underlies the 0th layer are sequentially determined to be settled just below the central part of the

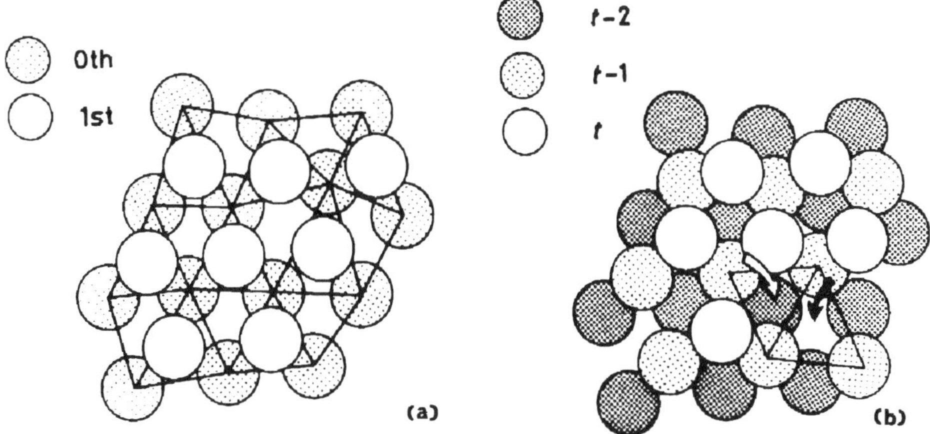

Fig. 3. Schematic drawings for explanation of computer simulation in the three-dimensional space. Both are views from below (from a basal layer). For easy understanding, spheres instead of points are used in the drawings. However, cells are assumed not to be spherical, but to be space filling polyhedra positioned at the centers of the spheres (see text). (a) Initial condition. (b) Positioning of points on layer t which are affected not only by the points on layer $t-1$ but also by the points on layer $t-2$.

alternate triangles on the 0th layer, so every other triangle on the 0th layer has a point below it (Fig. 3(a)). The computer simulation begins with the above-mentioned initial condition.

In general, for the purpose to determine a new point on the layer t, a triangle is successively selected at random among the triangles in the network on the upper layer (the layer $t-1$ which lies over the layer t). The layer $t-2$ affects the determination of a point on the layer t; if the triangle does not have just above it any point which belongs to the layer $t-2$ (which lies over the layer $t-1$) as shown by the solid arrow in Fig. 3(b), the triangle is used to determine a point of the present layer t so that a point is settled just below the central part of the triangle, and if the triangle has a point just above it (open arrow in Fig. 3(b)), no point is settled and the next triangle is again selected at random. This assumption seems reasonable; a new cell may jostle easier to the region indicated by the solid arrow because of no hindrance from a cell on the layer $t-2$, than the region indicated by the open arrow. When all triangles of the layer $t-1$ were tried, the process advances to a new layer. The process is summarized in Fig. 4. By repetition of such a process a pile of layers on which the points are distributed is obtained.

The actual computer simulation was performed as the above-mentioned scheme. When we used the inner center of the triangle for the central part of the triangle, the result is shown in Fig. 5. On the lower layers which are made after many processes, the points on every three layers are found to be approximately arranged on a vertical line which reminds us of an epidermal cell column.

For the purpose to present the result by two-dimensional display of cell boundaries,

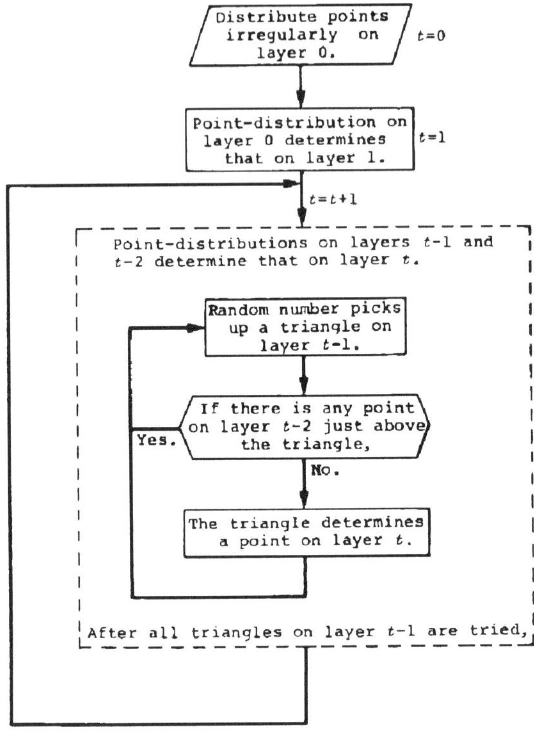

Fig. 4. Block diagram or the computer simulation of cell-column-formation in the three-dimensional space.

an approximation is used; since neighboring cell points do not exist on the same layer (Fig. 6), we had better consider three sequential layers as one unit. All points on the layers just above and below it ($t-1$ and $t+1$) are projected on the concerned layer t. Dirichlet domains are obtained from all points existing and projected on the layer t. Their boundaries are approximation to three-dimensionally interdigitating cell boundaries (Fig. 6). This approximated pattern will be called a pattern of the layer $[t-1, t, t+1]$. Patterns [0, 1, 2]~[3, 4, 5] and [11, 12, 13]~[14, 15, 16] are shown in Figs. 7(a) and (b). The lower patterns (large t-value) have become regular. Statistical distributions of area, edge length and exterior angles of Dirichlet polygons have become narrow. Histograms of edge length and their standard deviation are shown in Fig. 8. Similar results were obtained when using other series of random numbers. When the center of gravity of the triangle was used as the central part of the triangle, similar results were also obtained in the case of using the inner center of the triangle. However, when the center of the circumcircle of the triangle was used, the point distribution did not converge to make cell columns from such an initial condition as used in the former cases. Because, the center of the circumcircle is far from the central part of the triangle when an inner angle of the triangle is around or larger than a right angle.

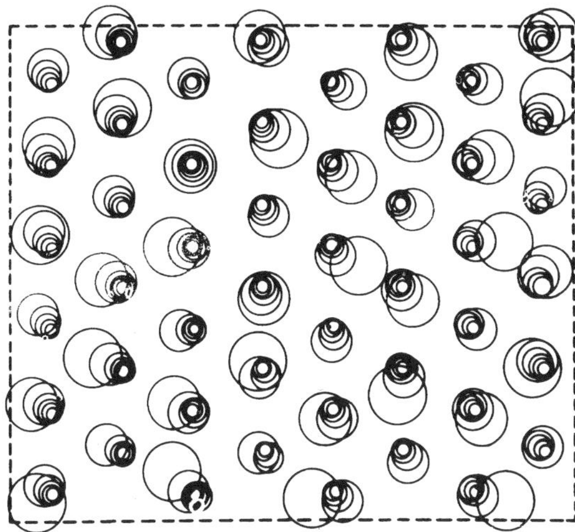

Fig. 5. Surface views of cell positions during computer simulation. On the 0th layer the sixteen points are irregularly distributed as described in Methods. The simulation process is performed using the inner center as the central part of the triangle. The results are displayed in superposition. Circles represent cell positions with their centers. The largest circles represent the cell positions on layer 0, the second represent those on layer 1, the third represent those on the layers 2, 3 and 4, ..., the smallest represent those on layers 14, 15 and 16. Broken line, the rectangular boundary (2: $\sqrt{3}$)where periodic boundary condition is used for computation.

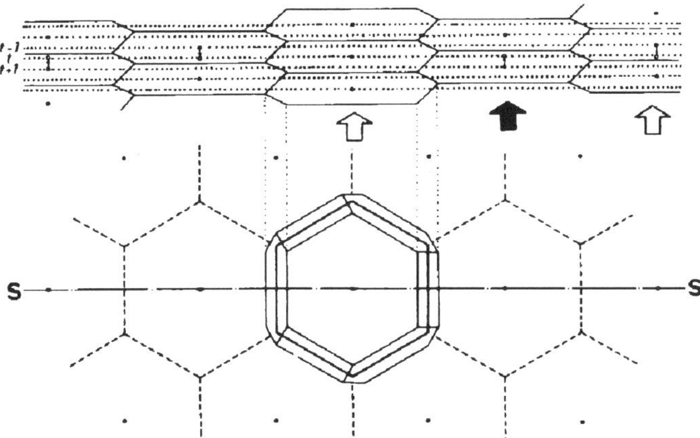

Fig. 6. Schematic representation or vertical section of a pile of layers (top) and its surface view (bottom). Neighboring cells are not on the same level, but show tri-level arrangement. Cells are interdigitating and stacked in a neat column normal to the surface. A cell which is a flattened tetrakaidecahedron is shown in a center of the bottom figure. This is approximately represented by a hexagon (thick solid line) which is a two-dimensional Dirichlet domain after projections of the neighboring points on the layer $t \pm 1$ onto the layer t (see text). Since the cell thickness is thinner in actual than that in the figure (the ratio of cell diameter to thickness is 10~15), the approximate representation of the surface view of the cell by a Dirichlet domain is closer to actual. SS, line where sectioned for the top figure.

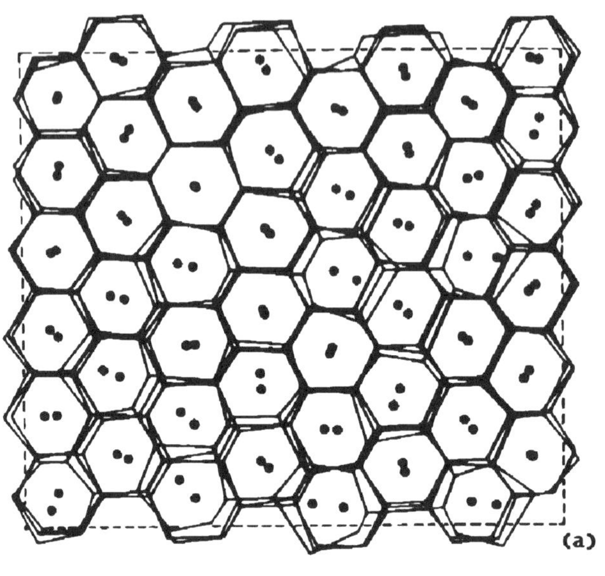

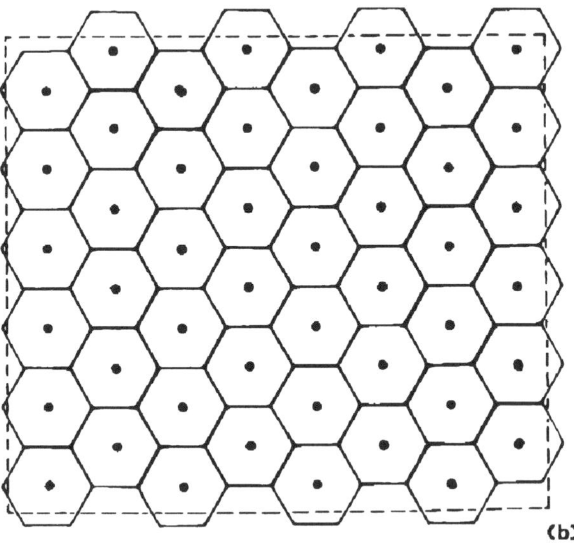

Fig. 7. Surface views of the formation of cell columns during computer simulation. Approximated cell boundaries are represented using Dirichlet polygons on the basis of the result in Fig. 5. Solid circle, the point which represents the cell and is used as Dirichlet center for cell boundary display. Cell patterns are displayed in superposition of the layers [0, 1, 2]~[3, 4, 5] (a) and the layers [11, 12, 13]~[14, 15, 16] (b).

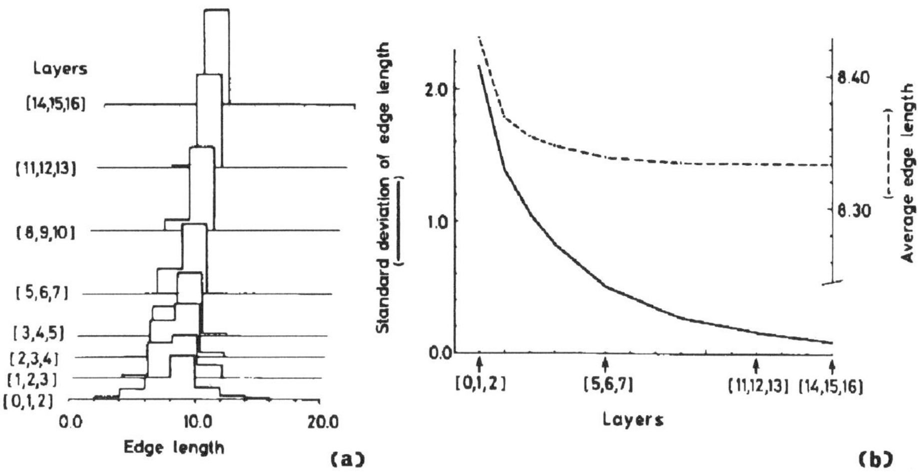

Fig. 8. Statistical analysis of distribution of edge length of Dirichlet polygons during computer simulation in Fig. 7. (a) Histograms of edge length of the patterns of the layers [0, 1, 2], [1, 2, 3], [2, 3, 4], [3, 4, 5], [5, 6, 7], [8, 9, 10], [11, 12, 13] and [14, 15, 16]. (b) Changes of average edge length (broken line) and standard deviation of edge length (solid line) during computer simulation.

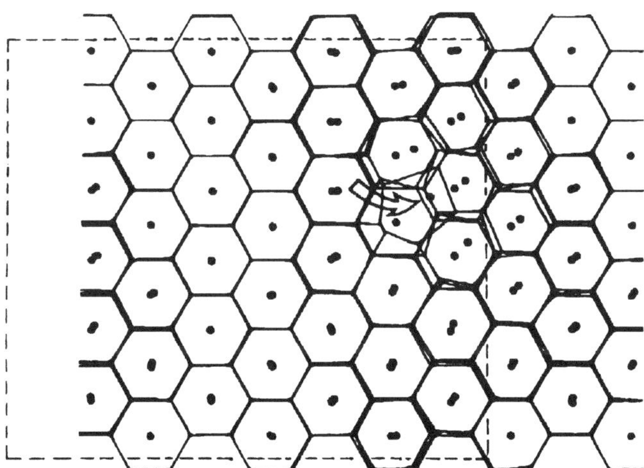

Fig. 9. Recovery of the regular cell arrangement. On the 0th layer, the sixteen points are arranged in a network of equilateral triangles and the one point (arrow) is displaced. The computation is performed as Fig. 5. The result is shown in a surface view superposing the layers [0, 1, 2]~[3, 4, 5].

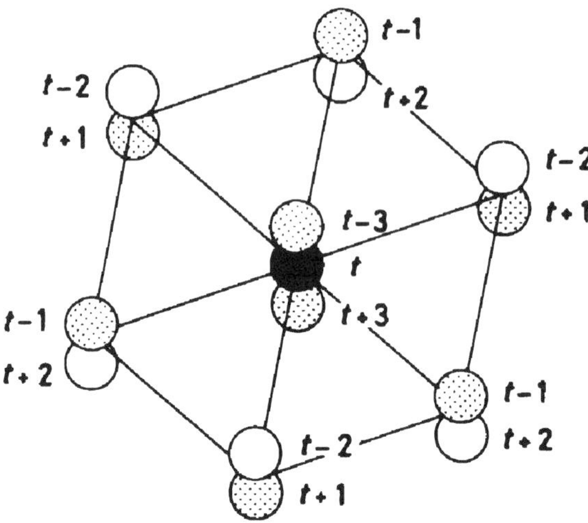

Fig. 10. The first and second nearest neighboring points (dotted spheres) and the third nearest (open spheres) around a point (solid sphere) on the layer t. The layer number to which a sphere belongs is respectively indicated. See text.

After the points were regularly arranged, one point was displaced. The regularity was recovered within several layers (Fig. 9).

The distribution of the points after many layers has become almost regular and we have shown it with the two-dimensionally approximated cellular boundaries by using the two-dimensional Dirichlet domains (Fig. 7(b)). However since the distribution of the points is three-dimensional, we will consider the three-dimensional Dirichlet domains here. From the regular distribution of the points, we could construct the three-dimensional Dirichlet domains which is the space-filling polyhedra and the same as tetrakaidecahedra topologically (Figs. 1 and 6); when the distance between the layers is small (Figs. 6 and 10), a point on the layer t has the two first nearest neighboring points on the layer $t \pm 3$, the six second nearest on the layers $t \pm 1$ (these neighbors correspond to the two large regular hexagonal faces and the six small hexagonal ones of the flattened tetrakaidecahedron as shown in Fig. 1), and the six third nearest on the layers $t \pm 2$ (these six correspond to the six square faces). The present computer simulation has demonstrated the establishment of epidermal cell columns which consist of flattened tetrakaidecahedra, and the similarity in the shape of all cells in a column (MENTON, 1976a). The simulated pattern should be considered after neglecting upper several layers because of a constant loss of surface cells in mammalian skin.

4. Discussion

The computer simulation has demonstrated the formation of regular vertical cell columns and their stability with the simple assumption that cells migrating upward from

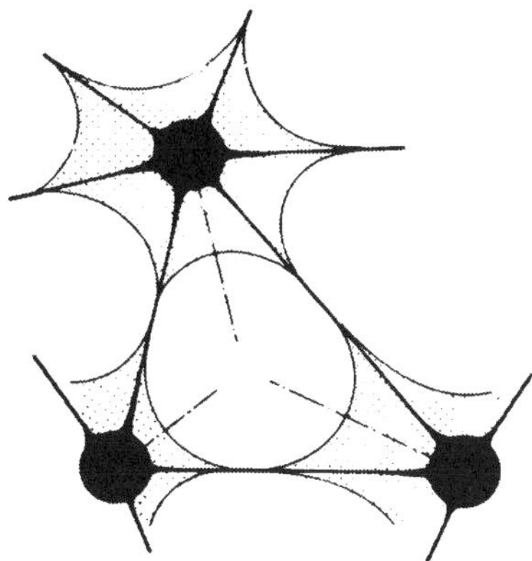

Fig. 11. A drawing explaining that a cell settles at the inner center of the triangle consisting of the upper cells. See text.

a basal layer jostle and settle at the less crowded area among the upper cells. The final shape of a cell in the simulation has been shown to be a flattened tetrakaidecahedra which was observed in the actual epidermis by MENTON (1976a). The architecture consisting of such cells is compatible with the organization and turnover of stacked cells.

We assumed that the less crowded area among the upper cells is the central part of the triangle consisting of the upper three cells. It may not be precisely where the central part is. The relatively small area around the central part of the triangle may be important, since the simulation has worked well with either case, the inner center or the center of gravity of the triangle. However there may be a plausibility in the case of the inner center, according to the following idea (Fig. 11): the neighboring cells on the upper layer are connected to each other to make triangles (solid lines) and spread gradually like webs (dotted areas). The web spreads symmetrically about a bisector line (chain lines) which divides equally the angle of the triangle, and the spreading speed is low when the angle is large and high when the angle is small. If so, the region around the inner center of the triangle may be less crowded and a new cell may settle here with high probability.

When rigid spheres are packed two-dimensionally and fixed on a flat plane, and when a new rigid sphere is put over them, the sphere should be positioned just above the center of the circumcircle of the triangle consisting of the lower three spheres according to solid geometry. Bearing such a geometrical fact in mind, the center of the circumcircle of the triangle was used as the central part of the triangle. The simulation did not work well as described in Results. This may indicate that cells in epidermis are inappropriately assumed to be such ones as rigid spheres.

The settlement of a new cell in the central part of the triangle may be affected by the turnover rate or the rate at which a new cell migrates upward. Because, epidermis lacking in cell columns is characterized by high mitotic activity in basal cells (CHRISTOPHERS *et al.*, 1974; MENTON, 1976a); regular cell stacking does not occur on the soles or palms (CHRISTOPHERS, 1971), and on healing skin in experimental wound or rubbing (CHRISTOPHERS, 1972). Also there is the finding that rapidly produced soap bubbles do not assemble into a columnar array during the experiment of cell column formation in a model system involving the association of randomly produced soap bubbles into a stable froth (MENTON, 1976b).

The settlement of a new cell among the upper cells takes place somewhere between cornified cell layers and a basal layer. We cannot indicate the specific layer in detail where the cell settles on the basis of the result of the simulation. However, we could say that cells in the higher layers than that where the cells settle, are fixed to each other or, at least, cells do not move actively by themselves, and that cells in the lower layers move actively. The bottom level of cell columns may be the boundary between the two regions where cell movements are active and inactive. Such a change of cell properties could be explained from cell differentiation, but it is not the purpose of the present investigation. Why cells are compressed normal to the skin surface is also another problem. However, the tri-level structure of epidermis may provide an answer to why the skin surface is parallel to a basal layer, when a new cell from below is assumed to fill the uppermost vacant position (solid arrow in Fig. 6) than the lower position of the bottom level of column array (open arrows in Fig. 6). Because, after such cell positioning, the bottom level of column array, even if it is rough, becomes approximately even.

The result of the computer simulation differs from the observations in two respects: (1) when viewed normal to the skin surface six-sided cells are predominant but there are few five- and seven-sided cells (Table 1 in MENTON, 1976a), and (2) most of six-sided cells are not exactly regular hexagons. These differences may be due to the simplification of the layers to which cells belong. Actual layers are not geometrically flat planes and are uneven everywhere. Vertical distance between layers is not strictly constant. This may be the reason of the second respect of the differences. And there may be some local insertions of an extra-layer between the complete layers, which makes five- or seven-sided polygons.

As a mathematical solution for division of space, without interstices, into uniform bodies of equal volume with an economy of surface in relation to volume, Kelvin in 1887 proposed an orthic tetrakaidecahedron which is the 14-sided body with eight hexagonal faces and six square faces and all the edges of the faces are equal. Since then, the extensive studies of "cell" shape in the three-dimensional space have been performed in vegetable parenchymas, foam of soap bubbles and leaden small-shots compressed by high pressure. These results showed generally that most "cells" were 14-sided bodies indeed, but unfortunately common shapes of the faces were pentagonal, not hexagonal nor square (see THOMPSON, 1942; LEWIS, 1943). The shapes of tetrakaidecahedra were not found except some rare cases (e.g. cork cells in LEWIS, 1928). However, MENTON (1976a) noticed in rare cases tetrakaidecahedral cells not only in cork cambium and the pith of woody plants, but also in mammalian skin. He said "This study has shown that cells of some tissues which are organized as stacked columns bear a remarkable similarity, whether they be plant or animal, and that such cells appear to be fundamentally of tetrakaidecahedral

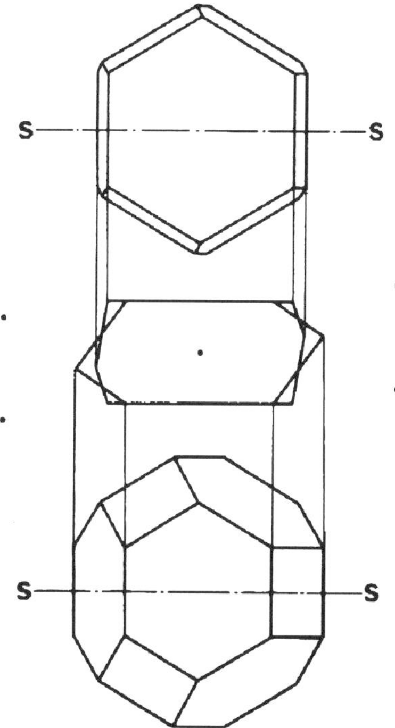

Fig. 12. The shape of a flattened tetrakaidecahedron is not unique even if the vertical thickness is fixed. Two cases are shown in the top and the bottom figures when viewed normal to the skin surface. The middle figure is vertical sections in superposition of the top and the bottom figures which are sectioned along lines SS respectively.

shape. In the case of epidermal cells, other intrinsic forces have modified this shape to produce a flattened form of the tetrakaidecahedron which is, none the less, a minimum-surface polygon".

The shape of a flattened tetrakaidecahedron is not unique even if the vertical distance (cell thickness) is fixed (Fig. 12). It is not definite how much adjacent cell columns are overlapped by each other. It might be determined so that the cells become minimum-surface polygons under other intrinsic forces as suggested by MENTON (1976a).

The present work has been carried out as a link in the chain of the investigations in histology and histogenesis with respect to individual cell properties by using Dirichlet description as a tool, which has been introduced in HONDA (1978).

Note added in proof. A related paper of a similar computer simulation under the initial condition of randomly arranged disks has been published entitled "Spontaneous architectural organization of mammalian epidermis from random cell packing" by H. Honda, M. Tanemura, and S. Imayama in *J. Invest. Derm.*, **106**, 312–315 (1996).

This work is partly supported by a Grant-in-aid for Cancer from the Japan Ministry of Education, Science and Culture. Acknowledgement is due to Mrs. Dorothy R. Smith in Harvard Forest, Harvard University for the typing of a manuscript, and Mrs M. Yamasaki-Koyama for drawings.

REFERENCES

ALLEN, T. D. and POTTEN, C. S. (1974) *J. Cell. Sci.*, **15**, 291.
ALLEN, T. D. and POTTEN, C. S. (1976) *Nature*, **264**, 545.
CHRISTOPHERS, E. (1971) *J. Invest. Derm.*, **56**, 165.
CHRISTOPHERS, E. (1972) *Virchows Arch. Abt. B Zellpath.*, **10**, 286.
CHRISTOPHERS, E., WOLFF, H. H. and LAURENCE, E. B. (1974) *J. Invest. Derm.*, **62**, 555.
HONDA, H. (1978) *J. Theor. Biol.*, **72**, 523.
LEWIS, F. T. (1928) *Science*, **68**, 625.
LEWIS, F. T. (1943) *Am. J. Bot.*, **30**, 74.
MACKENZIE, L. C. (1969) *Nature*, **222**, 881.
MACKENZIE, L. C. (1970) *Nature*, **226**, 653.
MACKENZIE. L. C. and LINDER, J. E. (1973) *J. Invest. Derm.*, **61**, 245.
MENTON, D. N. (1976a) *Am. J. Anat.*, **145**, 1.
MENTON, D. N. (1976b) *J. Invest. Derm.*, **66**, 283.
POTTEN, C. S. and ALLEN, T. D. (1975) *Differentiation*, **3** 161.
THOMPSON, D'Arcy W. (1942) *On Growth and Form*, 2nd ed., Cambridge University Press, London.

Kelvin Polyhedra and Analysis of Crystallization

M. TANEMURA

The Institute of Statistical Mathematics, 4-6-7 Minami-Azabu, Minato-ku, Tokyo 106, Japan

Keywords: (0 4 4 6)-Atom, (0 6 0 8)-Atom, Critical Nucleus, Face-Vector, Soft-Core Model

1. Introduction

The name of Lord Kelvin appears in many fields of science, as the name of unit of absolute temperature, as the name of thermodynamical work function, as the name of mathematical formula related with modified Bessel functions, as the name of variable transformation preserving harmonicity of a function, and so on.

However, the name Kelvin is also well known as to represent a specific type of polyhedron: truncated octahedron or *tetrakaidecahedron*. In reality, this polyhedron was already known to Archimedes, in the ancient age of Greece, and it was re-discovered by Russian crystallographer, Fedrov (WEYL, 1952).

But the importance of the conjecture by Lord Kelvin was that the tetrakaidecahedron would have a minimal surface area when it is tiled periodically into the three-dimensional space under the condition that the volume of the polyhedron is fixed. According to this reason, the tetrakaidecahedron is often called *Kelvin polyhedron*. Rigorously speaking, the Kelvin's tetrakaidecahedron has such a shape that its six square faces are flat and that its eight hexagonal faces are delicately curved. The detailed reasoning of Kelvin polyhedron may be discussed in the other papers of this special issue of *Forma*.

In this paper, however, we use the term "Kelvin polyhedron" as having all flat surfaces rather than including curved surfaces of its original name.

Kelvin polyhedron, in the form of its simplified terminology as above, appears in several fields. In crystallography, the Wigner-Seitz cell of body-centered cubic lattice is the truncated octahedron, i.e., Kelvin polyhedron. This is also equivalent to the first Brillouin zone of the face-centered cubic lattice.

In cell biology, the Kelvin polyhedron had been a typical model for the stacking tissue of cells, such as epidermis of some kind of mammalian species (in the latter case, however, the shape of cell is flattened although topologically equivalent to the Kelvin tetrakaidecahedron) (HONDA *et al.*, 1996).

In this paper, we will discuss the Kelvin polyhedra in connection with the analysis of atomic configuration of crystallization process. Since it is important to detect a crystalline nucleus from the atomic configurations of crystallization process, it is clarified the analysis should be based on the Voronoi polyhedra. It is thus useful to generally introduce the division of space into Voronoi polyhedra. This is presented in the next section. In Section 3, we introduce an index of polyhedron in order to characterize the topological shape of convex polyhedra. In Section 4, the computer simulations of crystallization process for the soft-core model are described. And then, it will be shown that the Kelvin polyhedron plays an important role for defining the crystalline nucleus. The results of analysis obtained by using this definition of nucleus are also discussed.

2. Tessellation of Space into Voronoi Polyhedra

For a given configuration $\{x_1, x_2, ..., x_N\}$ of N points (atoms) in a volume V of three-dimensional space, a Voronoi polyhedron Π_i of an atom i is defined as a set of all points of space which are closer to the atom i than to any atom j ($j \neq i$). The boundaries of Voronoi polyhedron consist of some of the perpendicular bisectors of line segments $x_{ij} \equiv x_j - x_i$, $j = 1, ..., N$ ($\neq i$). It is easily seen that all of Voronoi polyhedra are convex. The set of all Voronoi polyhedra for a given configuration of atoms fill the space without overlap and without gap uniquely. Thus, the set of Voronoi polyhedra constitutes a tessellation of space. The tessellation of space by Voronoi polyhedra is called a *Voronoi tessellation*. It will be recognized that Voronoi polyhedron is the Wigner-Seitz cell generalized to random configurations.

The geometrical structure of Π_i represents the local atomic configuration around the atom i. This fact will be seen as follows.

Let us define the atom, say j, to be *contiguous* to the atom i, if the perpendicular bisector of the line segment x_{ij} constitutes a face of Π_i. Then, the pair of atoms i and j which are contiguous to each other, is said to be a *contiguous pair*.

Denoting the total number of faces of Π_i by f, it is obvious that the number of atoms contiguous to the atom i is equal to f. In order to characterize the local configurations, further informations about the arrangement of atoms around the atom i are necessary. Among them, the number of edges of a certain face is informative.

Let us denote by F_{ij} the face of Π_i corresponding to the contiguous atom j to the atom i. Then, the number of edges of the face F_{ij} indicates the number of circumferential atoms which are contiguous to both i and j. In the following, we also represent the perpendicular bisecting plane of contiguous pair (i, j) by the same symbol F_{ij}.

Now, we pay attention to a certain edge of the face F_{ij}. And we assume the edge is common to a face, say F_{ik}, besides the face F_{ij}. Accordingly, this edge appears as a part of common line of two bisector planes F_{ij} and F_{ik}. Then, the atom k, which is just assumed to be contiguous to the atom i is also contiguous to the atom j. This is confirmed by the fact that the line common to both planes F_{ij} and F_{ik} is also common to the plane F_{jk}. We denote the common line by E_{ijk}. Thus, the three atoms i, j and k are contiguous with each other and are called a *contiguous triangle* (i, j, k), since the line segments of contiguous pairs (i, j), (j, k) and (i, k) constitute a triangle whose vertices are the atoms i, j and k. It will be obvious that the line E_{ijk} perpendicularly passes the center of circumscribing circle of the contiguous triangle (i, j, k). We denote also by E_{ijk} the edge we are thus far paying attention.

Next, we consider a vertex of the face F_{ij}. We can assume it is common to two edges E_{ijk} and, say E_{ijl}. Let us denote the vertex by V_{ijkl}. Then, by a similar argument as above, we can define a *contiguous tetrahedron* (i, j, k, l) whose vertices are the atoms i, j, k and l. It is also easily seen that the vertex V_{ijkl} is the center of circumscribing sphere of the contiguous tetrahedron (i, j, k, l). The tetrahedron is known as a *Delaunay tetrahedron*, and the set of all Delaunay tetrahedra for a given configuration of atoms is known to constitute another unique tessellation of space, Delaunay tessellation. The correspondence between the vertex of Voronoi polyhedron and the Delaunay tetrahedron indicates a *duality* between Voronoi and Delaunay tessellations. A computational algorithm for three-dimensional Voronoi tessellation based on this duality is given in TANEMURA *et al.* (1983).

From the discussions so far, it became clear that the geometry of Voronoi polyhedron is closely related to the local configuration of atoms which are connected with "contiguities". Faces of Voronoi polyhedron Π_i correspond to a set of atoms contiguous to the central

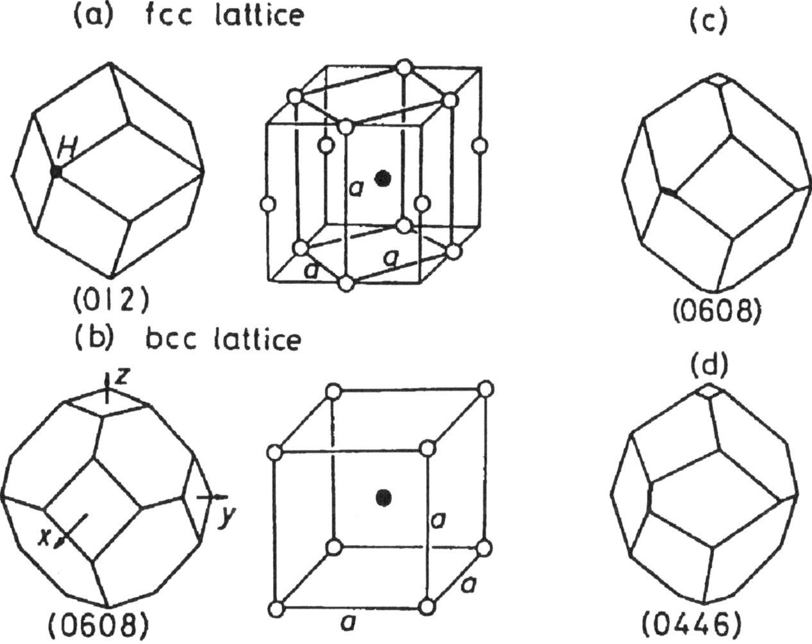

Fig. 1. Topological diagrams of fcc and bcc lattices and their Voronoi polyhedra. (a) fcc lattice, (b) bcc lattice. Black circles are the centered atoms of respective Voronoi polyhedra. If the parallelpiped indicated by bold lines in (a) is compressed from top and bottom, the Voronoi polyhedron (0 12) is transformed into (0 6 0 8) in (c). If the horizontal bold line in (c) is replaced by the vertical, (0 6 0 8)-polyhedron changes to the (0 4 4 6)-polyhedron as in (d).

atom i. Edges and vertices of each face F_{ij} of Π_i correspond to a set of atoms which are contiguous to the contiguous pair (i, j). Thus, the Voronoi polyhedron Π_i is determined by the set of all atoms contiguous to the atom i.

Let us illustrate the Voronoi polyhedron for the typical configuration of atoms (Fig. 1). Figures 1(a) and 1(b) represent the Voronoi polyhedron of an fcc (face-centered cubic) crystal and of a bcc (body-centered cubic) crystal, respectively. The Voronoi polyhedron of an fcc crystal is a rhombic dodecahedron, whereas that of a bcc crystal is a truncated octahedron (Kelvin polyhedron!). In these cases, each Voronoi polyhedron is of course equivalent to the Wigner-Seitz cell of corresponding lattice.

3. Indices of Polyhedra

In order to specify the shape of convex polyhedra, there are several ways of indexing them. Among them, we adopt the method of counting the number of faces of k-gons in such a way

$$(p_3, p_4, p_5, \ldots)$$

where p_k represents the number of k-gons of the polyhedron we concern.

This is called a simple *face-vector* (JENDROL', 1983) or a simple *p-vector* (BAYER and LEE, 1993) in the field of graph theory. Here, the adjective "simple" represents that all vertices of the polyhedron concerned has the degree three, in other words, that every vertex of a simple polyhedron is incident with three edges. Owing to this reason, the simple polyhedron is also called tri-valent (3-valent) polyhedron.

As regards the face-vector, the so-called Eberhard theorem holds (GRÜNBAUM, 1967):

Eberhard Theorem: There exists a three-dimensional polyhedron with p_k k-gons ($k \geq 3$, $k \neq 6$) and some number of hexagons if and only if

$$3p_3 + 2p_4 + p_5 = 12 + \sum_{k \geq 3} (k-6) p_k$$

is satisfied.

The above relation among the elements of face vector $(p_3, p_4, p_5, \ldots)$ is derived from the Euler's polyhedron theorem under the assumption that every vertex of the polyhedron is incident with three edges, i.e., the simplicity of polyhedron. This equation has some interesting features. If a polyhedron does not have faces whose number of edges is larger than six, then the equality $3p_3 + 2p_4 + p_5 = 12$ should hold. The above equation does not contain information about p_6. The values of p_6 that realizes a simple polyhedron are deeply considered (JENDROL', 1983). See also BAYER and LEE (1993).

We now show some examples of face-vectors. The face-vector of the Voronoi polyhedron of an fcc crystal in Fig. 1(a) is given by (0 12 0 0 0 0 ...) whereas that of a bcc lattice in Fig. 1(b) is (0 6 0 8 0 0 ...). If 0's continue infinitely, we omit these for simplicity and denote (0 6 0 8 0 0 ...), for instance, by (0 6 0 8).

We notice that the Voronoi polyhedron (0 12) for an fcc lattice is simple in the sense described as above, since at vertex H in Fig. 1(a) four edges meet. If the atoms are displaced from respective lattice points due to thermal oscillation, the vertex H is immediately decomposed into several vertices. On the other hand, the Voronoi polyhedron (0 6 0 8) for a bcc lattice is simple and is stable against the thermal oscillation of atoms.

The atom whose Voronoi polyhedron has a face-vector (p_3 p_4 p_5 ...) will be called (p_3 p_4 p_5 ...)-atom or (p_3 p_4 p_5 ...)-polyhedron. In the following, we often call the (0 6 0 8)-polyhedron also the Kelvin polyhedron because of their topological equivalence*.

4. Analysis of Crystallization Process

It is generally observed that, in the liquid state, the production rate of crystalline nuclei considerably increases, at certain range of temperature beneath the melting (freezing) point.

We suppose that we prepare a super-cooled fluid state, by repeating rapid cooling and compression processes, from an equilibrium fluid at a sub-melting point. Then, the following problems appear: Does this super-cooled state maintain as a metastable state? Or, does it show a crystallization to a solid state? The result will depend on the macroscopic conditions such as the rates of rapid cooling and conditions. And it is important to investigate the microscopic mechanisms underlying for this phenomena.

We have done the attempts, through computer simulation with molecular dynamics method, to realize the crystallization process.

4.1. Soft-core model and crystallization process

We choose the following pairwise interaction potential

$$\phi(r) = \varepsilon\left(\frac{\sigma}{r}\right)^n, \qquad \varepsilon > 0, \ \sigma > 0,$$

which is called the *soft-core model*. In our study, the system of the soft-core model with $n = 12$ was used.

It is known that any reduced quantity of the system of soft-core model becomes a function of a *reduced density* ρ^* defined by

$$\rho^* = \rho\left(\frac{\varepsilon}{kT}\right)^{1/4}$$

*To be legitimate, we can find two types of (0 6 0 8)-polyhedron. However, in the atomic configurations we are interested, another type of (0 6 0 8)-polyhedron which is topologically different from Kelvin polyhedron did not appear at all.

where $\rho = N\sigma^3/V$ is the number density.

Figure 2 shows the equation of state (the relation of compressibility factor PV/NkT to the reduced density ρ^*) of the soft-core model system at high density region. There are two branches in this system, amorphous state and crystalline state.

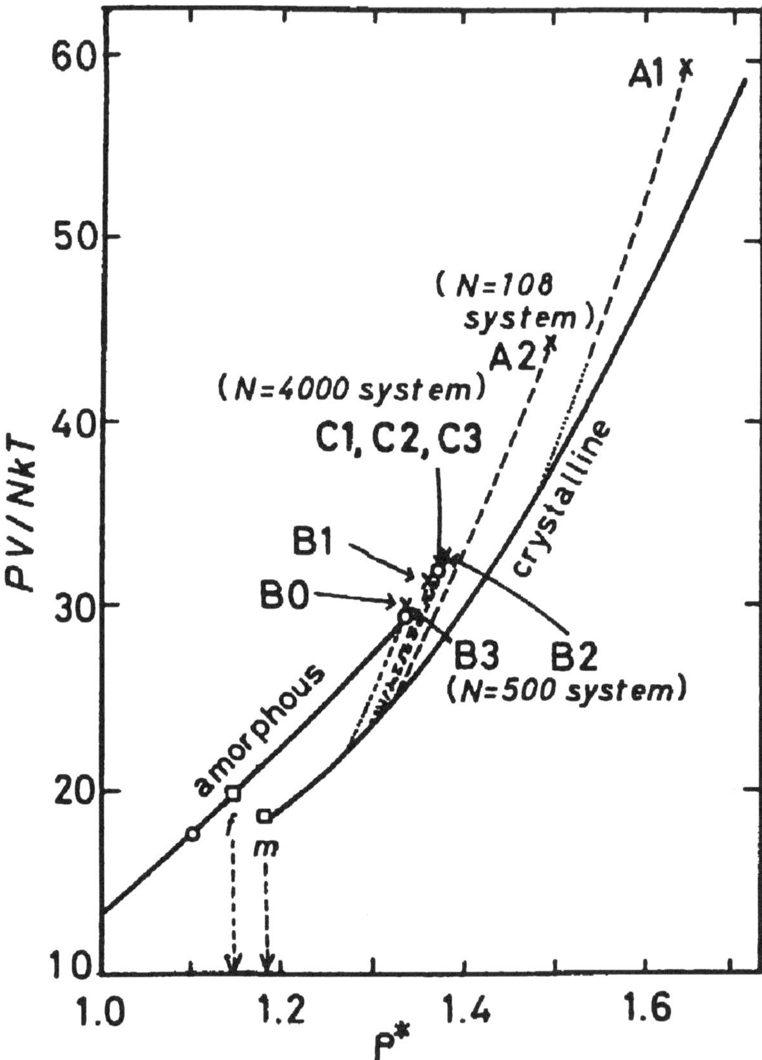

Fig. 2. Equation of state for the soft-core model system at high density region. Solid curves indicate amorphous and crystalline states, respectively. Each broken curve represents the relaxation path along which an unstable state moves under the total energy conservation law in the course of simulation run.

In this figure, the mark f represents the reduced freezing density $\rho^*_f = 1.15$ and the mark m the reduced melting density $\rho^*_m = 1.19$.

In our previous works (TANEMURA *et al.*, 1977, 1990), the molecular dynamics method of computer simulation was applied to the system of $N = 500$ and $N = 4000$. Simulation runs for $N = 500$ system are indicated as $B0$, $B1$, $B2$ and $B3$ in Fig. 2, whereas, for $N = 4000$ system, they are indicated by $C1$, $C2$ and $C3$. All simulations were performed under the cubic periodic boundary conditions. We do not present the parameter values used for the runs. For the details, see the references cited above.

Most of the simulation runs (except for $B0$) for $N = 500$ system exhibited the relaxation towards the crystalline state. A sample of time evolution of instantaneous compressibility factor $(PV/NkT)_t$ is shown in Fig. 3 for the run $B1$. In Fig. 3, the ordinate represents a "relaxation function" $\Phi(t)$ which is an instantaneous compressibility factor

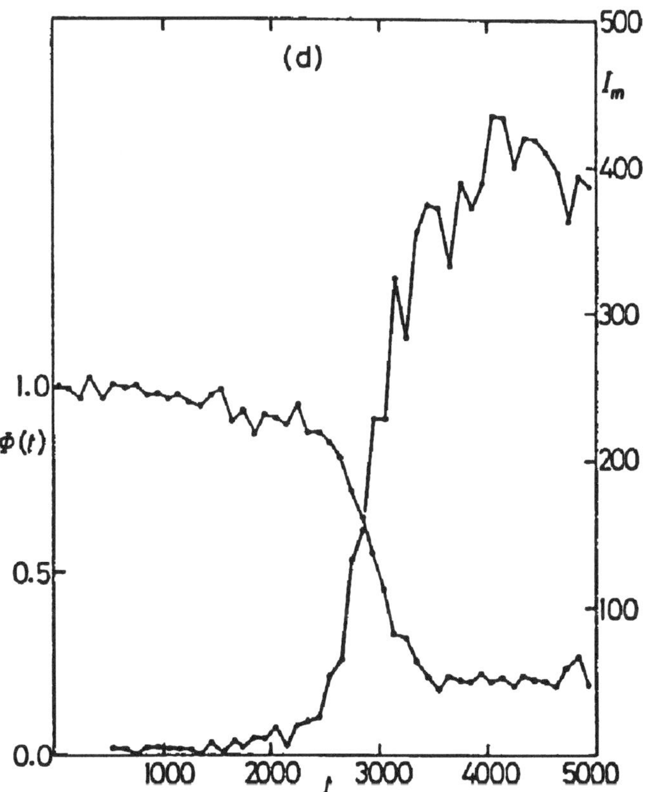

Fig. 3. Relaxation function $\Phi(t)$ and time evolution of the size of the main nucleus I_m for the run $B1$. Dots on the thin line denote the values of relaxation function averaged over 50 time steps. Bold line indicate I_m which is represented by the number of atoms shown on the scale at the right hand side.

transformed in such a way that $\Phi(0) = 1$ is satisfied and that $\Phi(\infty) = 0$ if the state of the system eventually tends to a crystalline state. These simulation runs $B1$, $B2$ and $B3$ strongly indicated the fact that the crystallization process happens for each of them.

By the method of analyzing the crystallization process, which is discussed in detail in the next subsection, it was revealed that the system of $N = 500$ atoms or less is too small to estimate, for example, a true critical nucleus size.

Accordingly, we have investigated the $N = 4000$ system, which is twice those of $N = 500$ system in linear scale. Three independent simulations were done for $N = 4000$ system which are indicated by $C1$, $C2$ and $C3$ in Fig. 2.

Figure 5 shows the time variation of instantaneous compressibility factor $(PV/NkT)_t$ for the runs $C1$ and $C2$. In all three runs, the decrease of $(PV/NkT)_t$ was observed, which indicates that the relaxation towards the crystalline state, namely, the crystallization has occurred from the unstable state.

4.2. *Crystalline nucleus and Kelvin polyhedra*

In order to examine the development of crystallization process in the computer simulation, we have to define a crystalline nucleus in the system. Although there might exist differences in local order and in local density between the nuclei and the remaining parts, the difference in local density may be small contrary to a rather large difference in density between gas and liquid in case of condensation. Then a method is required to detect the crystalline local order in a rather uniformly dense environment. For that purpose, it was found that the employment of Voronoi tessellation, discussed in Section 2, is most convenient (TANEMURA *et al.*, 1977).

In order to characterize atomic configurations, we used, in the following analysis, the configurations which were obtained by averaging atomic positions over 50 time steps. By

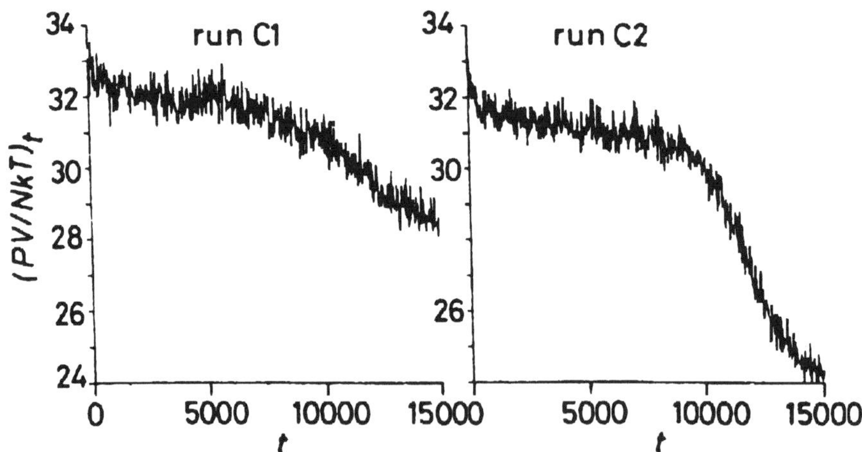

Fig. 4. Time variations of $(PV/NkT)_t$ for the runs $C1$ and $C2$ of $N = 4000$ atom system.

noting that a period of Einstein oscillation corresponds to a few ten time steps at densities we are concerned in our simulation, the configuration average over 50 time steps just above might be sufficient for eliminating the effect of thermal oscillations.

We performed the Voronoi tessellation for configurations of atoms thus obtained, and then made histograms of polyhedra classified by the indices or face-vectors mentioned previously. Figure 4 indicates the time variation of the histograms of face-vectors obtained for the run $B1$ of $N = 500$ system. From this figure, it can be seen that, as time goes on, the number of (0 4 4 6)-atoms first increases, then subsequently the number of (0 6 0 8)-atoms tends to increase and soon exceeds that of (0 4 4 6)-atoms. Finally (0 6 0 8)-atoms overwhelm the atoms of other classes. It was evident, by a separate

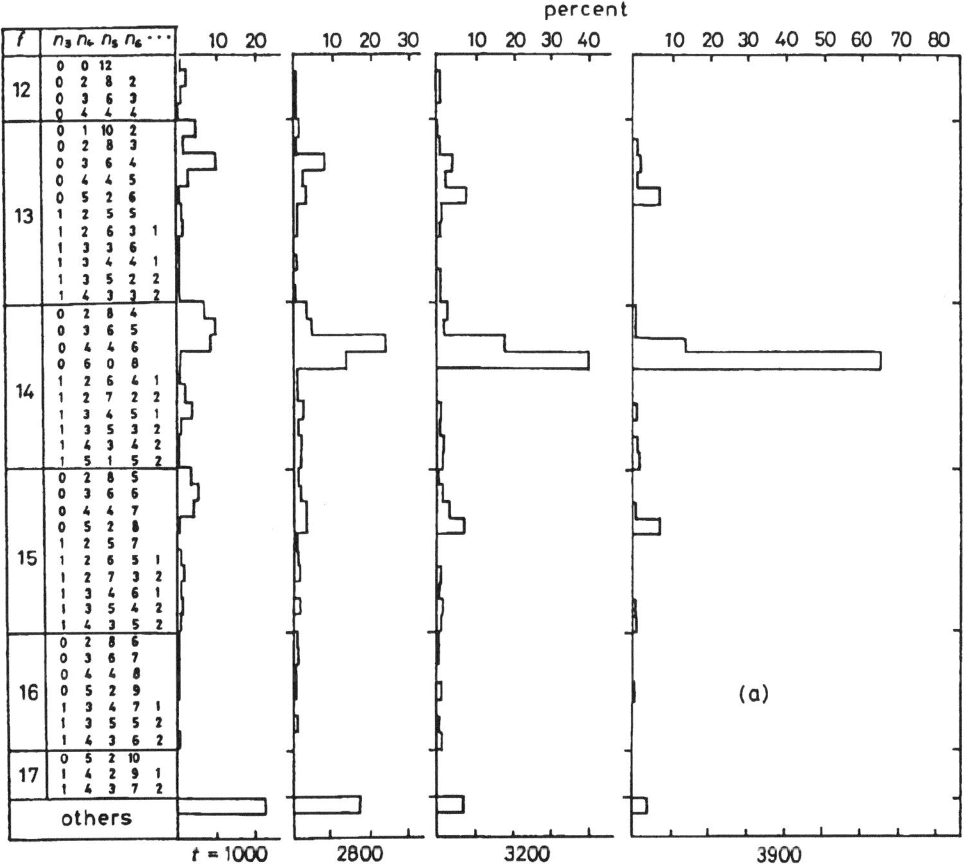

Fig. 5. Time evolution of histograms showing the fraction of Voronoi polyhedra classified according to face-vectors ($p_3\, p_4\, p_5$ ...) for the run $B1$. Polyhedra which are of minor abundance and hence not listed in the figure are gathered in the item "others".

analysis for a bcc solid, that the last histogram in Fig. 4 showed a close resemblance to that of bcc solid.

Similar qualitative features of histogram were obtained for other simulation runs *B*2, *B*3 of $N = 500$ system which also indicated relaxations to the crystalline state.

Owing to these facts, we can see that (0 6 0 8)- and (0 4 4 6)-polyhedra are expected to play an important role for the crystalline nuclear growth.

We noticed in Figs. 1(c) and 1(d) that the topological relation exists among these two polyhedra. If the bold vertical edge of the (0 4 4 6)-polyhedron transforms to a bold horizontal edge, a (0 4 4 6)-atom turns out to be a (0 6 0 8)-atom and vice versa. This transformation is caused by only a slight movement, that is, a thermal oscillation of contiguous atoms. Thus, a (0 4 4 6)-atom is topologically close to a (0 6 0 8)-atom.

We found, on the other hand, in the system of $N = 108$ atoms that the run A2 (see Fig. 2) showed a relaxation towards an isotropic fcc solid state. In this case, the competence among (0 4 4 6)-atoms and (0 6 0 8)-atoms was not observed. However, in this case, we can also realize the same situation as in the case of a bcc solid by compressing the system parallel to one of the axes of fcc crystalline lattice of the system so that it relaxes to a bcc solid with anisotropic distortion (TANEMURA *et al.*, 1978). This situation is partly illustrated in Figs. 1(a) and 1(b). If the fcc lattice is compressed from top and bottom along the (001) axis, namely, the *c*-axis indicated by the bold line in the diagram on the right hand side of Fig. 1(a), the (0 12)-polyhedron is transformed into (0 6 0 8)-polyhedron as indicated in Fig. 1(c).

Therefore, to both types of crystallization towards a bcc solid and an fcc solid, we may apply the same definition of a nucleus as follows:

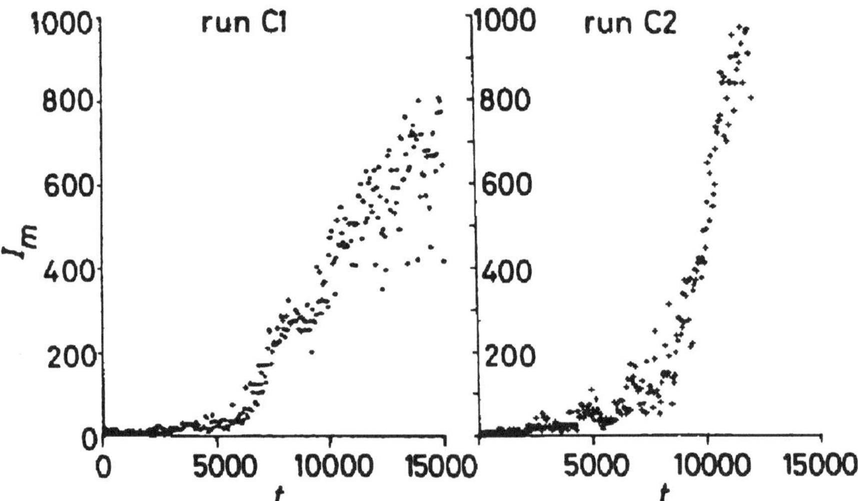

Fig. 6. Time variations of the size of the main nucleus I_m for both runs *C*1 and *C*2.

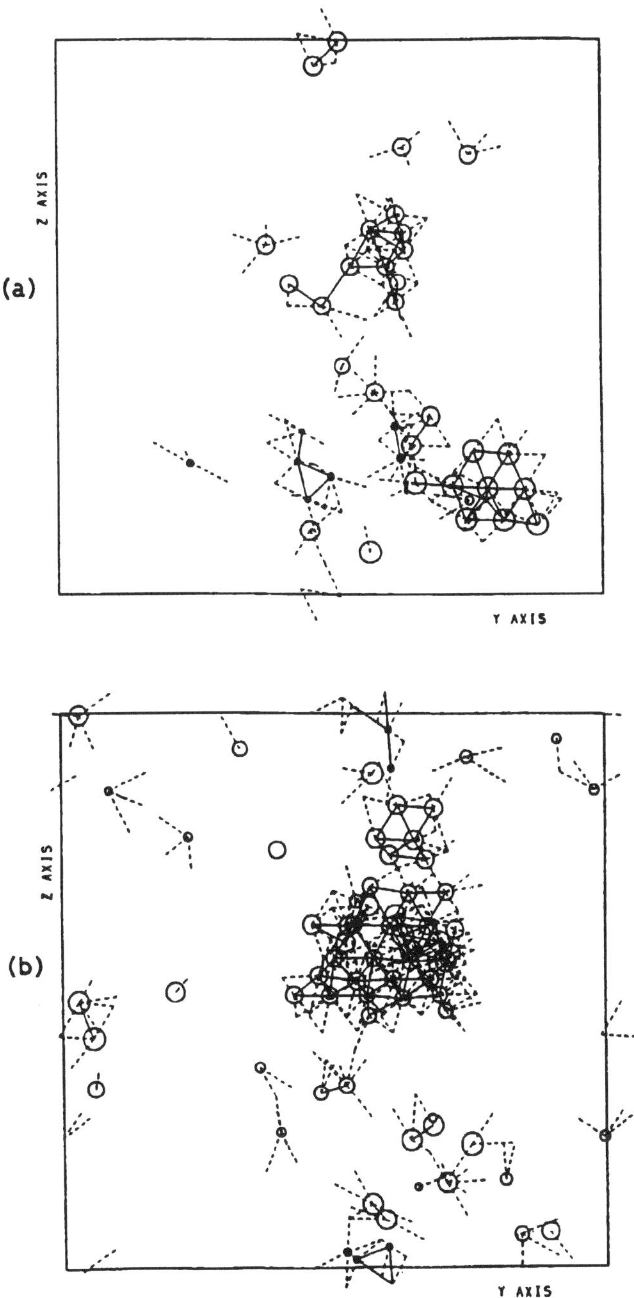

Fig. 7. XY-plotter diagrams of spatial distribution of (0 6 0 8)-atoms (solid circles) (0 4 4 6)-atoms (opposite end of every broken lines which are connected to solid circles) which form nuclei for the run $C2$. Each diagram shows a two-dimensional projection onto the YZ plane of three-dimensional distribution. (a) $t = 5800$. (b) $t = 6950$. These values of time step is those of ending time step of the average (see text).

Definition of crystalline nucleus

1. A nucleus contains at least (0 6 0 8)-atoms.
2. If (0 6 0 8)-atoms and/or (0 4 4 6)-atoms are contiguous to a (0 6 0 8)-atom, they are members of nucleus of the latter.
3. The number of the both kinds of atoms in a nucleus represents the nuclear size.

Using this definition of nucleus, we investigated the time variation of the size of nuclei. In Fig. 3, the time evolution of the largest nucleus I_m is presented by bold broken lines. The nucleus begins to grow gradually until its size becomes a few tens and then it grows rapidly. In the figure, it is obvious that the time dependence of transformed compressibility factor $\Phi(t)$ well correlates to the behavior of I_m.

The above definition made possible to visualize the spatial distribution of crystalline nuclei. By visualizing the nuclei for the crystallization process of $N = 500$ atom system, it was found that a nucleation often proceeds anisotropically, which might be due to the cubic periodic boundary condition employed during the simulation runs.

Then, this experience motivated the investigation of $N = 4000$ atom system. By utilizing the same definition of crystalline nucleus as above, the simulation runs $C1$, $C2$ and $C3$ were analyzed. Figure 6 is the time variation of I_m, correspondent with Fig. 4, for the runs $C1$ and $C2$. The behaviour of plots for the runs $C1$ and $C2$ show an excellent coincidence with that of $(PV/NkT)_t$ in Fig. 4.

For the curves of time variation I_m, we performed a statistical analysis to find the time when the change of slopes occurs and to obtain the size of nucleus at that time. Then, we estimated the mean value of the size of nucleus to be about 70~80. We can assume this value to be an estimate of the mean size of critical nucleus.

Figures 7(a) and 7(b) illustrate the spatial distribution of nuclei at early stage of crystallization for the run $C2$. These figures indicate the shape of nuclei is often relatively compact and spherical.

5. Discussion

In order to analyze the crystallization process, we found the importance of Kelvin polyhedra, i.e., (0 6 0 8)-polyhedra played in defining the crystalline nucleus. The reason why the crystallization process in the computer simulation results in favor of a bcc crystal is still not conclusive. In our analysis for $N = 500$ system, most of simulation runs which indicated the relaxation towards crystal phase showed indication of distorted bcc lattice structures. In the $N = 4000$ system, however, the degree of distortion in the nucleation is shown to decrease. Thus, we have a hope that a larger system of more than $N = 32000$ would show interesting features as regards the crystalline nucleation. In this situation also, the method of analysis based on Kelvin polyhedra would be again helpful.

The author would like to represent his thanks to Professors H. Matsuda, T. Ogawa, N. Ogita and A. Ueda for their continuing collaborations and discussions on the molecular dynamics studies of crystallization process. The author also represents his thanks to Prof. O. E. Barndorff-Nielsen,

Institute of Mathematics, Aarhus University, Denmark, for supplying the computer facilities for completing the manuscript during the author's stay at Aarhus University.

REFERENCES

BAYER, M. M. and LEE, C. W. (1993) Combinatorial aspects of convex polytopes, in *Handbook of Convex Geometry* (eds. P. M. Gruber and J. M. Wills), Elsevier Science Publishers, Amsterdam, Volume A, pp. 485–534.

GRÜNBAUM, B. (1967) *Convex Polytopes*, John Wiley & Sons, London.

HONDA, H., TANEMURA, M. and IMAYAMA, S. (1996) Spontaneous architectural organization of mammalian epidermis from random cell packing, *J. Invest. Dermatology*, **106**, No. 2, 312–315.

JENDROL', S. (1983) On the face-vectors of trivalent convex polyhedra, *Mathematica Slovaca*, **33**, 165–180.

TANEMURA, M., HIWATARI, Y., MATSUDA, H., OGAWA, T., OGITA, N. and UEDA, A. (1977) Geometrical analysis of crystallization of the soft-core model, *Prog. Theor. Phys.*, **58**, 1079–1095.

TANEMURA, M., HIWATARI, Y., MATSUDA, H., OGAWA, T., OGITA, N. and UEDA, A. (1978) Geometrical analysis of crystallization of the soft-core model in an FCC crystal formation, *Prog. Theor. Phys.*, **59**, 323–324.

TANEMURA, M., MATSUDA, H., OGAWA, T., OGITA, N. and UEDA, A. (1990) Molecular dynamics study of crystallization of the soft-core model, *J. Non-Cryst. Solids*, **117/118**, 883–886.

TANEMURA, M., OGAWA, T. and OGITA, N. (1983) A new algorithm for three-dimensional Voronoi tessellation, *J. Comput. Physics*, **51**, 191–207.

WEYL, H. (1952) *Symmetry*, Princeton University Press, Princeton, New Jersey.

Index